Genetics and Sports

Medicine and Sport Science

Vol. 54

Series Editors

J. Borms Brussels
M. Hebbelinck Brussels
A.P. Hills Brisbane
T. Noakes Cape Town

Genetics and Sports

Volume Editor

Malcolm Collins Cape Town

11 figures, and 12 tables, 2009

KARGER

Basel · Freiburg · Paris · London · New York · Bangalore ·
Bangkok · Shanghai · Singapore · Tokyo · Sydney

Medicine and Sport Science
Founded 1968 by E. Jokl, Lexington, Ky.

Malcolm Collins, PhD
MRC/UCT Research Unit for Exercise Science and Sports Medicine (ESSM)
South African Medical Research Council (MRC) and the
Department of Human Biology, University of Cape Town (UCT)
Sports Science Institute of South Africa
Boundary Road
Newlands 7700, Cape Town (South Africa)

This book was generously supported by **GATORADE**

Library of Congress Cataloging-in-Publication Data

Genetics and sports / volume editor, Malcolm Collins.
   p. ; cm. -- (Medicine and sport science, ISSN 0254-5020 ; v. 54)
  Includes bibliographical references and indexes.
  ISBN 978-3-8055-9027-3 (hard cover : alk. paper)
 1. Sports--Physiological aspects. 2. Human genetics. I. Collins,
Malcolm, 1965- II. Series: Medicine and sport science, v. 54. 0254-5020 ;
  [DNLM: 1. Genetic Phenomena. 2. Sports. 3. Athletic Injuries--genetics.
4. Athletic Performance--physiology. 5. Exercise--physiology. 6. Genetic
Techniques. W1 ME649Q v.54 2009 / QT 260 G3285 2009]
  RC1235.G46 2009
  599.93'5--dc22
                                                            2009024000

Bibliographic Indices. This publication is listed in bibliographic services, including Current Contents® and Index Medicus.

Disclaimer. The statements, opinions and data contained in this publication are solely those of the individual authors and contributors and not of the publisher and the editor(s). The appearance of advertisements in the book is not a warranty, endorsement, or approval of the products or services advertised or of their effectiveness, quality or safety. The publisher and the editor(s) disclaim responsibility for any injury to persons or property resulting from any ideas, methods, instructions or products referred to in the content or advertisements.

Drug Dosage. The authors and the publisher have exerted every effort to ensure that drug selection and dosage set forth in this text are in accord with current recommendations and practice at the time of publication. However, in view of ongoing research, changes in government regulations, and the constant flow of information relating to drug therapy and drug reactions, the reader is urged to check the package insert for each drug for any change in indications and dosage and for added warnings and precautions. This is particularly important when the recommended agent is a new and/or infrequently employed drug.

All rights reserved. No part of this publication may be translated into other languages, reproduced or utilized in any form or by any means electronic or mechanical, including photocopying, recording, microcopying, or by any information storage and retrieval system, without permission in writing from the publisher.

© Copyright 2009 by S. Karger AG, P.O. Box, CH–4009 Basel (Switzerland)
www.karger.com
Printed in Switzerland on acid-free and non-aging paper (ISO 9706) by Reinhardt Druck, Basel
ISSN 0254–5020
ISBN 978–3–8055–9027–3
e-ISBN 978–3–8055–9028–0

Contents

VII Preface
Collins, M. (Cape Town)

1 Key Concepts in Human Genetics: Understanding the Complex Phenotype
Gibson, W.T. (Vancouver, B.C.)

11 Nature versus Nurture in Determining Athletic Ability
Brutsaert, T.D. (Albany, N.Y.); Parra, E.J. (Mississauga Ont.)

28 Genetics and Sports: An Overview of the Pre-Molecular Biology Era
Peeters, M.W. (Leuven/Brussels); Thomis, M.A.I.; Beunen, G.P. (Leuven);
Malina, R.M. (Austin, Tex./Stephenville, Tex.)

43 Genes, Athlete Status and Training – An Overview
Ahmetov, I.I. (St. Petersburg/Moscow); Rogozkin, V.A. (St. Petersburg)

72 Angiotensin-Converting Enzyme, Renin-Angiotensin System and Human Performance
Woods, D. (Newcastle upon Tyne)

88 α-Actinin-3 and Performance
Yang, N.; Garton, F.; North, K. (Sydney, NSW)

102 East African Runners: Their Genetics, Lifestyle and Athletic Prowess
Onywera, V.O. (Nairobi)

110 Gene-Lifestyle Interactions and Their Consequences on Human Health
Pomeroy, J. (Phoenix, Ariz./Umeå); Söderberg, A.M.; Franks, P.W. (Umeå)

136 Genetic Risk Factors for Musculoskeletal Soft Tissue Injuries
Collins, M. (Cape Town); Raleigh, S.M. (Northampton)

150 Innovative Strategies for Treatment of Soft Tissue Injuries in Human and Animal Athletes
Hoffmann, A.; Gross, G. (Braunschweig)

166 Gene Doping: Possibilities and Practicalities
Wells, D.J. (London)

176 Genetic Testing of Athletes
Williams, A.G. (Alsager); Wackerhage, H. (Aberdeen)

187 The Future of Genetic Research in Exercise Science and Sports Medicine
Trent, R.J.; Yu, B. (Camperdown, NSW)

196 Author Index

197 Subject Index

Preface

Athletic performance and the occurrence of sports-related injuries are both multifactorial conditions which are determined by the complex and poorly understood interactions of both environmental and genetic factors. Although much work has been done to identify the non-genetic components that are associated with performance and susceptibility to injuries, there is an ever growing body of research investigating the genetic contribution to these phenotypes. This book, which contains contributions by a broad range of scientist and clinicians from several disciplines – including, but not limited to, human molecular genetics, clinical genetics and exercise science – covers a number of topics in an attempt to obtain an integrated and holistic understanding of the field. The recent sequencing of the human genome and development of several tools and methodologies have successfully been used to investigate the genetic contribution to many complex diseases. This area of research has more recently been applied to the fields of exercise science and sports medicine. This publication reviews past, current and future applications, as well as the ethical concerns, of genetic research in the fields of exercise science and sports medicine.

The introductory chapter of this book highlights current and key concepts in human genetics, in particular with respect to its application to understanding multifactorial conditions. This is followed by a chapter exploring the often misunderstood relationship between nature (genetics) and nurture (common environmental effects such as training, diet, etc.) in determining athletic ability. The third chapter summarises the methodologies initially used during the pre-molecular biology era and the estimates obtained for gene and environmental contributions for performance-related phenotypes. The molecular genetics of performance is specifically investigated in the subsequent four chapters. These chapters include a review of the specific DNA sequence variants currently believed to be associated with athletic performance and response to training, critical reviews of the roles of the much investigated variants within the angiotensin-converting enzyme and α-actinin-3 genes and performance, and finally an exploration of the genetic and lifestyle of the East African runners who have dominated distance running events for the last half century.

This book also contains a chapter exploring the interaction of lifestyle, such as physical active or sedentary, with genetic make-up and its implications on human health, in particular understanding mechanisms underlying specific diseases, such as obesity, and its prevention. This is followed by a chapter summarising the recent developments in the identification of genetic risk factors for musculoskeletal soft tissue injuries.

The current and possible application of gene therapy as well as other novel therapeutic strategies such as stem cell and growth factor therapies in injured athletes is also reviewed. This is followed by a chapter exploring the potential abuse of genetic information and technologies, together with other developments in molecular biology, (gene doping) to enhance performance. The ethical framework in which genetic research is done and its application is not straightforward. Exercise scientists and sports physicians need to keep abreast of advances in medical ethics broadly and customise them efficiently to the field. These issues are discussed in the second last chapter, which is followed by a review on the new technologies and paradigms which will influence future genetic research in exercise science and sports medicine.

Malcolm Collins, Cape Town

Key Concepts in Human Genetics: Understanding the Complex Phenotype

William T. Gibson

Department of Medical Genetics, Child and Family Research Institute, University of British Columbia, Vancouver, B.C., Canada

Abstract

The recent sequencing of a reference human genome has generated a large number of DNA-based tools, which are being used to locate genes that contribute to disease. These tools have also enabled studies of the genetics of non-disease traits such as athletic fitness. Sport scientists should keep in mind three major factors when designing such studies and interpreting the literature. First of all, the methods used to assign a biological trait (be it performance related or disease related) to a specific gene are not as powerful as is commonly believed. Second, the methods used are thought to be more robust for disease-related traits than for normal physical characteristics, likely because there are many more biological factors contributing to the latter. Third, additional levels of variability continue to be uncovered in the human genome; these may ultimately contribute more to physical differences between human beings than the levels studied over the past decade. This introductory chapter will aim to equip the reader with the necessary vocabulary to understand and interpret genetic studies targeted to sport fitness and sport-related injury.   Copyright © 2009 S. Karger AG, Basel

Key Genetic Concepts and Terms

Since the earliest days of competitive sport, athletes and their trainers have known that athletic performance is modifiable by dedicated training. However, some individuals appear to be naturally gifted with athletic ability, such that their performance prior to training is above average, and their performance after training is consistently excellent. Often encapsulated as the 'nature vs. nurture debate', the degree to which athletic ability is determined by inherited physical characteristics, as opposed to training-and diet-based regimens, has excited much interest.

The identification of genes that encode discrete biological molecules opened a new avenue of investigation – that of correlating physical traits with differences between individuals in DNA sequence. In many cases, it is easy to see how a physical trait such as height might impact on athletic performance: longer legs allow for greater distance to be covered with each stride. Increased cardiac output or increased lung capacity (relative to overall body size) would similarly be predicted to enhance athletic performance. The rapid growth in tools for DNA analysis has allowed researchers to identify an increasingly diverse number of genes and genetic markers, and in turn to correlate these with physical characteristics.

The biological information necessary to form all tissues of the human body is contained in the fertilized egg. Most of this information is physically located in the nucleus of the cell, though some is contained within energy-processing mitochondria. As the fertilized egg divides, the DNA that carries this information is copied and redistributed throughout the new cells as they are made. The term gene is used to refer to a discrete sequence of DNA that produces a biologically active product. If different sequences of the same gene coexist in a population, that gene is said to have different forms termed alleles. An allele is a variant sequence of an identified gene. Persons with different alleles for a particular gene are said to be heterozygous, those with identical alleles on both chromosomes are homozygous. The genome is the sum total of all genes and connecting sequences present in an individual or a population. The term 'genome' can be used to refer to one unique series of genes (haploid genome), to the redundant series of genes present in an organism (diploid genome), or to the collection of genes present in a large number of individuals under study (population or species genome). Such distinctions are important, because it is often assumed that information from population studies may be directly applied back to individuals. This may or may not be the case, as will be discussed later.

DNA is made up of a combination of four chemical bases, referred to by the letters A, C, G and T. A string of such letters makes up a DNA sequence, such as AATGCG. By default, the most common combination of letters found in a particular region of the genome is taken to be the 'normal' sequence. Differences from that order of nucleotides are considered to be DNA variants. Many variants are possible. In the example above, the sequence AAGGCG would be considered a variant, because it differs at the third position from the left: a G exists where a T would normally be found. Variants that change one chemical letter are referred to as point mutations or single nucleotide polymorphisms (SNPs), depending on their frequency in the population and whether they have a major effect on health. If the AATGCG sequence were present in 55% of the sequences in a study population, and the AAGGCG sequence were present in 45%, the 'T' variant would be considered the 'normal variant,' and the 'G' variant would be considered a SNP. If, on the other hand, 99.95% of sequences studied were found to be AATGCG and 0.05% were found to be AAGGCG, the 'G' would be considered a rare variant and termed a 'mutation'. Debate exists whether the term 'mutation' should be reserved exclusively for when the 'mutant' DNA sequence has major effects on physical characteristics, or perhaps for when the variant can be proven to have arisen at a definable time. For rare disorders wherein a DNA sequence variant is 100% associated with disease, the use of 'mutation' is well-recognized. In human population studies wherein the function of the DNA sequence may not be clear, the term 'mutation' is often used if the DNA variant is present only rarely in the population (i.e. at a frequency of 1% or less). In this chapter, no attempt will be made to settle this debate, and both the terms 'rare variant' and 'mutation' will be used, because both are found frequently in the literature.

The genotype of an individual refers to his or her specific DNA sequence for a known gene or DNA marker. By a similar derivation, the phenotype is a specific physical trait in an individual. Either term may be used narrowly to describe an individual sequence (e.g. AATGCG) or trait (e.g. total lung volume). Either term may also be used broadly, to refer to the sum total of all genetic or physical traits measured in a study. A haplotype (haploid genotype) refers to a known series of DNA sequences that occur on the same chromosome, and thus on the same strand of DNA. Haplotypes are often used in genetic studies, and may or may not contain specific genes. Haplotypes may contain a combination of SNPs, point mutations and other DNA sequences.

Mendelian traits are physical traits that obey Gregor Mendel's laws of inheritance, and hence are encoded by distinct genes. Genetic diseases are typically referred to as mendelian, but it is important to recognize that most physical traits associated with athletic performance will not be mendelian traits. Mendelian traits are transmitted with high frequency through families, which means that the risk to an immediate relative (parent, sibling or offspring) is rather high, on the order of 25–50%. Dominant traits are passed directly from parent to child; one allele is inherited from one parent who also has the trait, be it a healthy or a disease-related trait. Siblings and offspring of someone with a dominant trait have a 50% risk of inheriting the same allele, and hence the same trait. Recessive traits manifest in the children of two carrier parents; one mutant allele must be inherited from each (unaffected) parent, leading to a 25% risk to siblings. A special exception exists for sex-linked/X-linked recessive traits; these manifest in 50% of male siblings, with a 25% risk to all siblings when males and females are considered together. For mendelian traits, the physical trait correlates very highly with the susceptibility allele; often, 100% of individuals who inherit the mutant gene(s) will manifest the trait. It is also possible for an allele to be codominant, wherein heterozygotes manifest a phenotype that is intermediate between either of the two possible homozygote phenotypes. Tautological though it may be to say, complex traits are not so simple. Complex traits are physical characteristics that are influenced by both genetic and environmental factors such as diet, intrauterine environment and activity level. Complex traits are often quantitative traits, and in the normal population these terms are largely equivalent. Quantitative traits are physical characteristics for which a range of measurable values are possible, and are usually considered to be continuous variables. Quantitative traits include height, weight, forced vital capacity, stroke volume, etc. By contrast, qualitative traits are discrete physical characteristics that are either present or absent. Most complex traits encountered in sport science are assumed to be quantitative traits.

Genetic mapping is the process of assigning biological traits to genes, and/or of assigning genes to specific regions of chromosomes. Chromosomes are packages of

DNA that are visible during cell division. Because chromosomes are easily studied, they have become a convenient handle for gene mapping. Genes are mapped onto a chromosomal space much like a postal address might be noted on a cartographer's chart. A distinction should be made between genes (defined above) and DNA markers. DNA markers (also known as genetic markers or sequence tags) are used as reference points to describe the location of genes relative to each other. Just as a landmark on a map can be any feature of significance (a bridge, hill, church, pub, intersection, etc.), a DNA marker is merely a nucleotide sequence of known chromosomal location: it need not have an identifiable RNA or protein product. SNPs are a commonly used type of DNA marker, in part because a large number have been identified. SNPs are often assigned an 'rs' (reference SNP) number, such as rs1815739. It is easy to determine whether an individual carries a particular SNP, and to determine how many copies of that SNP he or she has. For the above reasons, SNPs are well suited for population studies. Apart from their use as markers, most SNPs do not have a known biological function.

A locus is a region of a chromosome that has a physical trait, DNA marker, allele or gene associated with it. A locus may be a theoretical entity and not necessarily a gene, though a genetic locus is often assumed to contain a gene with functional relevance to the physical trait under study. The locus itself may be a DNA sequence that controls the activity of one or more genes (e.g. by binding an activator protein). The locus can also be a 'neighbour' to a nearby sequence that is the actual determinant of the trait. If epidemiological data associate a physical trait such as height with a particular chromosome, that region is said to be a 'locus' for height [1]. It may be helpful to think of a locus as a low-resolution hypothesis: once data are sufficient to associate a physical trait with a region of the genome, a locus is defined and the relevant gene(s) or other sequences in the region are sought using higher-precision methods. A locus is a type of genetic 'neighbourhood', used as a term of reference until the actual 'address' is found in terms of the specific DNA sequence. Additional information, concepts and excellent diagrams may be found in the review by Attia et al. [2].

Genome-Wide and Candidate Gene Association Studies

The recent sequencing of the human genome has generated an extraordinary number of tools for DNA analysis, including several million SNPs [3]. Notable successes in the identification of rare mendelian single-gene disease traits have spurred widespread enthusiasm for DNA analysis as a predictive factor for disease risk. The rapid decline in the cost of genotyping these SNPs has spurred an equally rapid growth in studies that attempt to determine the contribution of genetic variation at the level of the DNA sequence to physical variation at the level of the whole organism. Analysis of such contributions is termed genotype-phenotype correlation. Though the usefulness of DNA analysis for mendelian genetic disorders is undeniable, the usefulness of DNA analysis for most complex traits remains to be clarified.

An association study is an epidemiological tool that measures the degree to which an identifiable factor accounts for the risk for a particular disease. The factor may be exposure to an environmental agent (e.g. cigarette smoking) or 'exposure' to carrier status for a genetic trait (e.g. a DNA marker near the myoglobin gene) [3]. A genetic association study correlates an identifiable genetic trait with risk for disease or 'risk' for improved health. Typically, these studies either employ a candidate gene approach or a genome-wide approach. Studies that select a specific gene or genes and correlate variation in these genes with a disease or physical trait are termed candidate gene association studies. The genes in question are considered as candidates based on their known function, presumed to be relevant to the disease under study. Studies that examine DNA markers at many positions on multiple chromosomes are termed genome-wide association studies (GWAS). A GWAS will employ a large number of DNA markers that are scattered throughout all chromosomes. Each of these markers will be studied ('typed,' short for 'genotyped') in a population of individuals, and these study participants will also be assessed for their physical characteristics (i.e. their phenotype). The ultimate goal of both candidate gene and GWAS is to establish genotype-phenotype correlations. A genotype-phenotype correlation is an association between a DNA sequence and a physical trait. The ultimate association may be strong or weak. Single-gene mendelian traits are examples of strong genotype-phenotype correlations; a child who tests positive for two copies of the delta-F508 mutation in the *CFTR* gene is extremely likely to develop cystic fibrosis later in life [4], and specific mutations in human myostatin correlate strongly with increased muscle mass [5]. However, it is important to remember that most genotype-phenotype correlations for nondisease traits are rather weak. Red hair and green eyes are often seen in individuals who carry polymorphisms at the *MC1R* gene, but DNA variation at this locus does not guarantee that someone will develop these traits, nor is *MC1R* the only gene that controls hair and eye pigmentation [6].

Candidate gene association studies are less expensive than GWAS, because fewer markers must be genotyped. Thus, candidate gene studies are frequently performed to assess the risk of sport-related injury that might be imparted by a particular DNA variant [7]. A SNP in one of the collagen genes could be genotyped among swimmers with rotator cuff injuries and among swimmers without such injuries, to assess the risk conferred by carrier status for that SNP. Equally well, candidate gene studies could be performed to assess the 'risk' for high performance [8]. A SNP in or near the gene for hemoglobin might be genotyped in a number of endurance runners, and the frequency of this SNP among high-performance endurance runners (the 'cases') compared with members of the general population (the 'controls'). Several such studies have been done, but sport scientists should interpret reported findings with caution.

Though candidate gene studies are inexpensive and relatively straightforward to do, the initial associations that they discover are usually not found to be true when studied in more detail. There are a number of reasons for this high false discovery rate (reviewed elsewhere [9]); it can be difficult to eliminate all possible sources of bias in a candidate gene association study. A major source of bias in many studies is population genetic substructure: the historical mixing between human populations that differed in the prevalences of genetic and physical characteristics. Suffice it to say that the actual ancestry of most individuals is much more complex than can be identified through questioning by a researcher, and this fact can lead to spurious statistical associations between ancient genetic markers and contemporary physical traits [10]. Other issues such as potential genotyping errors, blinded testing and reporting bias are also worth considering, but are not dealt with in detail here.

Definitive proof that a candidate gene (or variant in such a gene) is truly predictive of risk for sports injury would require a long-term prospective study in which DNA samples were collected in a large cohort of individuals, who were then followed over time until a sufficient number of them had developed the injury. Since this type of study is tremendously time consuming and expensive, an intermediate standard of proof is typically considered acceptable, wherein the findings of the preliminary association study are replicated in a larger population. For example, an association between SNPs in collagen genes and rotator cuff injury described in 100 cases and 250 controls from one academic center in a large city might be considered to be 'replicated' if, on assessment on a nationwide basis, carriers of the at-risk variant were found to have a higher prevalence of rotator cuff injuries in the previous 10 years than noncarriers. Typically, however, when such replication studies are attempted, the previous statistical association is no longer seen. Part of the difficulty likely lies in the fact that it is easy to come up with a biologically plausible hypothesis about how a particular DNA variant *might* influence disease. Whether the variants that *actually exist* in the human population *truly do* influence disease is another matter entirely. The answer is unknown at the time the study is done, but the accumulated experience from a number of candidate gene association studies for a wide variety of biological traits suggests that this method is prone to a high false discovery rate (i.e. the positive predictive value is low) [9, 11].

GWAS are thought to be more robust to replication than candidate gene studies. The former are often touted as being 'agnostic' or 'hypothesis-free,' since no prior assumptions are made about the specific gene or genes involved in susceptibility to disease (or high performance). However, GWAS are not, in fact, hypothesis-free. The hypothesis they test is whether there is biological variation in the human genome that is common enough to confer sufficient population-attributable risk to be detectable by the methods used. In simpler terms, GWAS test the hypothesis that there is, in fact, an association *somewhere in the genome*.[12] GWAS have achieved notable successes with biological traits for which genetic susceptibility is believed to be relatively high [13]. However, like candidate gene association studies, the loci identified in GWAS often are not replicated when larger studies are done, or when attempts are made to narrow the locus down to a specific gene [9]. In this context, it is particularly important to remember that most genotype-phenotype correlations for nondisease traits are rather weak. A corollary to this is that, for most performance-related traits and for most sport-related injuries, evidence for genetic contributions should be weighed carefully and replicated in large independent populations before being accepted as valid [14]. Because of the large number of SNPs genotyped, GWAS perform a number of comparisons that is several orders of magnitude greater than most epidemiological studies; thus, it is appropriate to seek much lower p values (on the order of 5×10^{-8} instead of 0.05) [15, 16].

SNPs vs. Copy Number Variants: New Discoveries, New Technologies and New Challenges

GWAS also fail to assess all of the variability in the human genome. It has recently been discovered that a large amount of genetic differences between individuals are due neither to mutations nor to SNPs, but to duplications and deletions of genes or groups of genes [17, 18]. The groups of extra and/or missing DNA sequences on human chromosomes have been termed copy number variants (CNVs) [19], and reveal a previously uncharacterized architecture of the genome. Thus, the size of a normal individual's genome may differ substantially from that of other normal people, and from the theoretical 'reference human genome.' The regions that differ may contain entire genes or groups of genes, such that people may vary not only in their sequence at specific detectable SNPs, but also in the number of genes that they have. Most of these CNVs are transmitted stably through families and do not cause genetic disease. It may be helpful to think of the presence of CNVs as representing a certain amount of 'wobble' in the size of the human genome.

The global significance of CNVs for normal physiology is not clear, but CNVs are thought to hold significant potential as a source of phenotypic variation. Indeed, the impact of CNVs on human physical characteristics may well be greater than the impact of SNPs, either at the individual or population levels. This is a new and exciting area of research; whether CNVs impact athletic performance or susceptibility to sport-related injury remains to be evaluated. Estimates vary, but each person probably has approximately 12 of these CNVs (6 per haploid genome), and roughly half of the CNVs found so far encompass one or more genes [19]. The impact of a particular insertion or deletion CNV on physical characteristics depends on the gene(s) involved, which correlates with the size of the CNV and the gene density of the region where it occurs. At the time of this writing, nearly 20,000 human CNVs have been described, at just over 6,000 genomic sites (http://projects.tcag.ca/variation, accessed 16 Jan 2009). A relevant feature of these polymorphisms is that CNVs tend to be found outside of functioning genes rather than within them. The fact that gene-rich areas are relatively depleted of CNVs suggests that CNVs occurring in or near genes tend to have a mildly harmful effect on those genes. If CNVs had no such effect, they would likely be distributed more randomly throughout the genome. The corollary of this phenomenon is that CNVs that are found in, near or including genes are expected to have a higher likelihood of having a biological effect than a SNP (possible exceptions include SNPs that result in an amino acid change in the protein product of a gene).

The Implications of Human Genomic Structure for Sport Science

Sport science researchers who are contemplating genetic studies that target traits such as high performance or injury susceptibility should be aware of the following properties of the human genome. The proportion of the genome that is made up of functioning genes is believed to be less than 3% [20]. The average coding length of a human gene is approximately 1,500 bp or 1.5 kb [21]. Estimates of the frequency of SNPs in the human genome vary from between 1 in every 180th nucleotide pair to 1 in every 279th nucleotide pair, on average [22, 23]. Thus, it is likely that polymorphic SNPs will be present in or near any candidate gene(s) selected for study. The proportion of the genome that is susceptible to copy number variation is currently believed to be on the order of 12% [24], an estimate that is likely to rise as more types of CNVs are described and better methods are applied to their detection. Thus, study designs should not focus exclusively on SNPs in plausible candidate genes – they should incorporate methods that assess copy number variation of the gene of interest. Similarly, GWAS should incorporate assessment of CNVs throughout the whole genome of the study participants. Copy number variants may be uncovered through the use of technologies like microarray comparative genomic hybridization ('array-CGH') and representational oligonucleotide microarray analysis (ROMA) and paired-end mapping, techniques that compare an individual's genome to a reference genome made up of an 'average' of several individuals' DNA [19, 25]. Several microarrays are available on the market [26]; these arrays have different levels of resolution, which translates into the ability to detect insertions and deletions of different sizes. Detection accuracy may also be improved by using multiple independent computational approaches to analysis of the data [27].

In addition to corroborating data from in vitro functional studies (such as expression levels for each identified allele), validation of any association findings will require replication in larger cohorts, ideally in one or more independent populations, and preferably with a prospective study design. Such stringent criteria are difficult to achieve, particularly among performance athletes where large sample sizes may not be available. At the very least, large control cohorts should be sampled from the same ethnocultural population as the athletes themselves, such that adequately narrow confidence intervals are available for the 'background' frequencies of the DNA variants studied. Population substructure also may dramatically skew the results of association studies [10]. Markers of population ancestry exist and should be genotyped where possible, particularly in populations where admixture is known to have occurred historically [28].

One major strength that sport science may draw upon is the fact that numerous tools are available to characterize the phenotype of performance athletes in detail, and to define the precise nature of the pathology in injured athletes. Detailed phenotyping, properly applied, may improve the power to detect a biological effect in a small cohort (relative to a large cohort that is studied in less detail). Similarly, detailed genotyping that incorporates DNA markers of population ancestry and assessment of copy number variants should be the goal of studies designed to correlate performance traits with human genes.

References

1 Weedon MN, Lettre G, Freathy RM, et al: A common variant of HMGA2 is associated with adult and childhood height in the general population. Nat Genet 2007;39:1245–1250.
2 Attia J, Ioannidis JP, Thakkinstian A, McEvoy M, Scott RJ, Minelli C, Thompson J, Infante-Rivard C, Guyatt G: How to use an article about genetic association. A. Background concepts. JAMA 2009;301: 74–81.
3 Altshuler D, Daly MJ, Lander ES: Genetic mapping in human disease. Science 2008;322:881–888.
4 Rowntree RK, Harris A: The phenotypic consequences of CFTR mutations. Ann Hum Genet 2003; 67:471–485.
5 Schuelke M, Wagner KR, Stolz LE, Hubner C, Riebel T, Komen W, Braun T, Tobin JF, Lee SJ: Myostatin mutation associated with gross muscle hypertrophy in a child. N Engl J Med 2004;350:2682–2688.
6 Sulem P, Gudbjartsson DF, Stacey SN, et al: Genetic determinants of hair, eye and skin pigmentation in Europeans. Nat Genet 2007;39:1443–1452.
7 September AV, Schwellnus MP, Collins M: Tendon and ligament injuries: the genetic component. Br J Sports Med 2007;41:241–246; discussion 246.
8 Bray MS, Hagberg JM, Perusse L, Rankinen T, Roth SM, Wolfarth B, Bouchard C: The human gene map for performance and health-related fitness phenotypes: the 2006–2007 update. Med Sci Sports Exerc 2009;41:35–73.
9 Attia J, Ioannidis JP, Thakkinstian A, McEvoy M, Scott RJ, Minelli C, Thompson J, Infante-Rivard C, Guyatt G: How to use an article about genetic association. B. Are the results of the study valid? JAMA 2009;301:191–197.
10 Marchini J, Cardon LR, Phillips MS, Donnelly P: The effects of human population structure on large genetic association studies. Nat Genet 2004;36:512–517.
11 Prince JA, Feuk L, Sawyer SL, Gottfries J, Ricksten A, Nagga K, Bogdanovic N, Blennow K, Brookes AJ: Lack of replication of association findings in complex disease: an analysis of 15 polymorphisms in prior candidate genes for sporadic alzheimer's disease. Eur J Hum Genet 2001;9:437–444.
12 Ioannidis JP: Non-replication and inconsistency in the genome-wide association setting. Hum Hered 2007;64:203–213.
13 Willer CJ, Speliotes EK, Loos RJ, et al: Six new loci associated with body mass index highlight a neuronal influence on body weight regulation. Nat Genet 2009;41:25–34.
14 Chanock SJ, Manolio T, Boehnke M, et al: Replicating genotype-phenotype associations. Nature 2007;447:655–660.
15 Hoggart CJ, Clark TG, De Iorio M, Whittaker JC, Balding DJ: Genome-wide significance for dense SNP and resequencing data. Genet Epidemiol 2008; 32:179–185.
16 McCarthy MI, Abecasis GR, Cardon LR, Goldstein DB, Little J, Ioannidis JP, Hirschhorn JN: Genome-wide association studies for complex traits: consensus, uncertainty and challenges. Nat Rev Genet 2008;9:356–369.
17 Iafrate AJ, Feuk L, Rivera MN, Listewnik ML, Donahoe PK, Qi Y, Scherer SW, Lee C: Detection of large-scale variation in the human genome. Nat Genet 2004;36:949–951.
18 Sebat J, Lakshmi B, Troge J, Alexander J, Young J, Lundin P, Maner S, Massa H, Walker M, Chi M, Navin N, Lucito R, Healy J, Hicks J, Ye K, Reiner A, Gilliam TC, Trask B, Patterson N, Zetterberg A, Wigler M: Large-scale copy number polymorphism in the human genome. Science 2004;305:525–528.
19 Feuk L, Carson AR, Scherer SW: Structural variation in the human genome. Nat Rev Genet 2006; 7:85–97.

20 Feuk L, Marshall CR, Wintle RF, Scherer SW: Structural variants: changing the landscape of chromosomes and design of disease studies. Hum Mol Genet 2006;15:R57–R66.
21 Eyre-Walker A, Keightley PD: High genomic deleterious mutation rates in hominids. Nature 1999;397:344–347.
22 Crawford DC, Akey DT, Nickerson DA: The patterns of natural variation in human genes. Annu Rev Genomics Hum Genet 2005;6:287–312.
23 International HC: A haplotype map of the human genome. Nature 2005;437:1299–1320.
24 Redon R, Ishikawa S, Fitch KR, et al: Global variation in copy number in the human genome. Nature 2006;444:444–454.
25 Korbel JO, Urban AE, Affourtit JP, Godwin B, Grubert F, Simons JF, Kim PM, Palejev D, Carriero NJ, Du L, Taillon BE, Chen Z, Tanzer A, Saunders AC, Chi J, Yang F, Carter NP, Hurles ME, Weissman SM, Harkins TT, Gerstein MB, Egholm M, Snyder M: Paired-end mapping reveals extensive structural variation in the human genome. Science 2007;318:420–426.
26 Carter NP: Methods and strategies for analyzing copy number variation using DNA microarrays. Nat Genet 2007;39:S16–S21.
27 Baross A, Delaney AD, Li HI, Nayar T, Flibotte S, Qian H, Chan SY, Asano J, Ally A, Cao M, Birch P, Brown-John M, Fernandes N, Go A, Kennedy G, Langlois S, Eydoux P, Friedman JM, Marra MA: Assessment of algorithms for high throughput detection of genomic copy number variation in oligonucleotide microarray data. BMC Bioinformatics 2007;8:368.
28 Tian C, Gregersen PK, Seldin MF: Accounting for ancestry: population substructure and genome-wide association studies. Hum Mol Genet 2008;17:R143–R150.

William T. Gibson, MD, PhD, FRCPC, FCCMG
Dept. of Medical Genetics, University of British Columbia, Child and Family Research Institute
950 West 28th Avenue, Room A4–182
Vancouver, BC V5Z 4H4 (Canada)
Tel. +1 604 875 2000, ext. 6783, Fax +1 604 875 2373, E-Mail wgibson@cw.bc.ca

Nature versus Nurture in Determining Athletic Ability

Tom D. Brutsaert[a] · Esteban J. Parra[b]

[a]Department of Anthropology, The University at Albany, SUNY, Albany, N.Y., USA, and [b]Department of Anthropology, University of Toronto Mississauga, Mississauga Ont., Canada

Abstract
This chapter provides an overview of the truism that both nature and nurture determine human athletic ability. The major thesis developed is that environmental effects work through the process of growth and development and interact with an individual's genetic background to produce a specific adult phenotype, i.e. an athletic or nonathletic phenotype. On the nature side (genetics), a brief historical review is provided with emphasis on several areas that are likely to command future attention including the rise of genome-wide association as a mapping strategy, the problem of false positives using association approaches, as well as the relatively unknown effects of gene-gene interaction (epistasis), gene-environment interaction, and genome structure on complex trait variance. On the nurture side (environment), common environmental effects such as training-level and sports nutrition are largely ignored in favor of developmental environmental effects that are channeled through growth and development processes. Developmental effects are difficult to distinguish from genetic effects as phenotypic plasticity in response to early life environmental perturbation can produce lasting effects into adulthood. In this regard, the fetal programming (FP) hypothesis is reviewed in some detail as FP provides an excellent example of how developmental effects work and also interact with genetics. In general, FP has well-documented effects on adult body composition and the risk for adult chronic disease, but there is emerging evidence that FP affects human athletic performance as well.

Copyright © 2009 S. Karger AG, Basel

Human athletic ability is determined by numerous sociocultural, psychological, anatomic, and physiological factors. The biological factors, such as an individual's aerobic capacity, are frequently termed complex traits, i.e. traits with a multifactorial genetic and environmental etiology. Gifted athletes are clearly a collection of many complex traits and so the athletic phenotype must arise from of a fortuitous combination of many genes and many environmental exposures. Some environmental effects are obvious and conceptually straightforward, like training. For example, it would be inconceivable to win the Tour de France without a years-long program of physical preparation. Other environmental effects, perhaps a majority, are indirect and poorly understood. For example, it is unknown whether vigorous physical activity in childhood is a necessary precursor for the development of an elite athletic phenotype in adulthood [1, 2]. On the genetic side, major gene effects are also conceptually straightforward, but most genetic effects are small and/or indirect reflecting gene-gene interaction (epistasis) or gene-environment (GE) interaction. Interaction effects are frequently not trivial, as demonstrated in recent studies of epistasis and physical activity traits in mice [3]. Further, it should also be considered that many elite athletes are the product of GE correlation which occurs when an individual self-selects for, or is urged to participate in, a sporting discipline for which they show an early aptitude. Thus, a complete explanation to account for the full range of human variation in athletic performance remains a considerable challenge.

This chapter will take a broad view and will present evidence that both nature (genetics) and nurture (environment) determine the variance in human athletic ability. Because many of the other papers in this volume focus on specific genes and sport performance, this chapter will instead provide a historical perspective vis-à-vis genetic research with some emphasis on the limitations of current approaches and areas for future investigation. For environmental effects, this chapter will focus on 'developmental effects' defined as environmental effects that are channeled through the process of growth and development leading to irreversible changes on the adult phenotype. Developmental effects have emerged in the public health literature as an important causal mechanism with the power to explain some of the variance in adult chronic disease risk. Thus, the chapter will explore the thesis that such effects could also have a major impact on human athletic performance.

Genetics and Athletic Ability

Multiple lines of evidence going back more than one century leave little doubt that genes influence athletic ability [for reviews, see 4–10]. Quantitative genetic studies of twins, families, or extended pedigrees yield a population summary statistic termed *heritability* (h^2), where heritability in the *broad sense* refers to the proportion of phenotypic variance attributable to all genetic effects. Numerous studies [reviewed in 9] have demonstrated a significant h^2 for traits relevant to athletic performance. These include studies of aerobic capacity and the response to training [11–13], anaerobic performance [14, 15], muscle strength and power [16–18], neuromuscular coordination [19], bone density [20], body size and composition [21–26], muscle fiber type distributions [27–31], cardiovascular variables including blood pressure [32–37], blood lipid biochemistry [38], blood glucose, insulin, and peripheral insulin sensitivity [24, 39–41], substrate utilization and metabolic rate [42–44], pulmonary function [45–49], and hormones and hormonal response to training [50–52]. One frequently cited study is that by Bouchard et al. [11, 12] revealing significant familial aggregation for VO_{2max} and VO_{2max} trainability for individuals in the sedentary state, after controlling for age, sex, body mass, and body composition. Heritability for both of these traits was near 50%, suggesting a substantial genetic contribution to baseline aerobic capacity and the response to training.

Unfortunately, heritability studies do not reveal the specific genetic architecture of a complex trait, i.e. these studies cannot: (1) distinguish between polygenic and/or major gene effects; (2) provide information on allele frequencies; (3) describe the mode of inheritance; (4) localize the chromosomal locations of genes, or (5) identify the gene products involved. To identify specific genes, two major approaches are widely applied: family-based linkage analysis and gene-association studies. Family-based linkage analysis is conducted on related individuals and the approach is based on the co-segregation of a putative (causal) trait locus with a known marker locus. Linkage implies the close proximity on the chromosome of causal and marker loci. Linkage analysis has a long history of success in mapping loci for simple monogenetic (mendelian) diseases such as Huntington's disease [53], but generally the approach only works if the causal locus has a high penetrance and a large effect on the trait in question. Linkage analysis may not have adequate statistical power and mapping resolution to detect genes of modest effect [54]. This is an important issue for understanding the variance in human athletic performance as elite athletes likely emerge on a favorable genetic background where individual alleles are both common in the general population and have only modest effects on the phenotype. In any case, there is a small literature reporting results of family-based linkage analyses with direct relevance to athletic performance. Numerous loci have been linked to traits which include VO_{2max} and VO_{2max} response to training [55–57], stroke volume, cardiac output, and blood pressure during exercise [58–62], training-induced changes in body composition [63–66], and insulin and glucose response to habitual physical activity or exercise [67].

Association studies include candidate gene approaches and various association mapping strategies that are conducted at the level of the whole genome. Candidate gene studies have proliferated in the last two decades since early studies in the 1970–1980s demonstrated no association of human physical performance traits with the classical blood group polymorphisms [9]. In a recent iteration of the Human Gene Map for Physical Performance Phenotypes [68], candidate gene studies comprised the majority (>90%) of the total literature. Interestingly, a sizable proportion of that literature (~15%) was focused on one specific gene polymorphism, the angiotensin converting enzyme (ACE) insertion/deletion polymorphism. Despite intensive focus, the significance of the ACE gene, in a genetic causal sense, remains controversial due to the inherent limitations of association studies. The candidate gene approach requires large sample sizes to identify loci with small effects. Additionally, the approach is susceptible to the problem of false association due to chance, bias, or confounding [69]. This is true both for case-control studies, i.e. comparing allele frequencies in elite athletes with the general population, and in studies that evaluate mean phenotypic values by genotype in a homogenous study group.

Genome-wide association (GWA) strategies rely on linkage disequilibrium to detect associations between traits and gene markers. GWA is a powerful and efficient means to dissect the genetic basis of complex traits [54, 70], and the advent of high-density genotyping arrays has allowed a shift away from candidate gene studies to GWA. Using GWA, there have been many recent successes in the elucidation of genes involved in disease processes such as type 2 diabetes [71–75], breast cancer [76–80], and prostate cancer [81–85]. Two key elements of these successes have been the collection of large sample sizes (i.e. thousands of individuals) with the consequent increase in study power to detect loci of modest effect, and the exponential advances in genotyping technologies that have dramatically improved genome coverage. To date, GWA has not been applied explicitly to study exercise traits, although some traits, such as body composition, may be relevant to human performance. Like candidate gene studies, GWA can be hindered by population stratification that can lead to an increase in false positives [86–89].

Another potential complicating issue that future genetic studies of human performance will need to consider is that there is extensive structural variation in the human genome [90–95]. Here, the term structural variation broadly refers to genomic alterations involving DNA segments larger than 1 kb [94, 96]. This term includes balanced changes (e.g. inversions, translocations) and those that alter copy number (copy number variants or CNVs). Of particular interest is the large number of CNVs described in humans. As of September 2008, more than 5,000 CNV loci have been catalogued in the Database of Genomic Variants (available at http://projects.tcag.ca/variation/). Not surprisingly, there have been recent studies investigating the role of CNVs in human phenotypic variation. In addition to the well-known influence of CNVs on sporadic and mendelian diseases [96, 97], recent research has indicated that CNVs have an important impact on levels of gene expression [98], and also play a role in complex traits and diseases, such as Crohn's disease [99], systemic lupus erythematosus [100], psoriasis [101], autism [102, 103], and schizophrenia [104]. Interestingly, there is also evidence indicating that CNVs may be involved in adaptation to different environments. A recent study has shown that individuals from populations with high-starch diets have, on average, more copies of the salivary amylase (AMY1) gene than those with traditionally low-starch diets [105]. Salivary amylase is responsible for starch hydrolysis, and AMY1 copy number is correlated positively with salivary amylase protein level. It has been proposed that having higher AMY1 copy numbers improves the digestion of starchy foods and for this reason AMY1 copy number has been subject to natural selection in human populations. As a working hypothesis, CNVs could affect human athletic performance particularly if variation exists in pathways that are directly involved in exercise, e.g. ATP synthesis, modulators of the cardiovascular response to exercise like ACE, and so forth.

As mentioned above, recent GWA efforts have identified hundreds of associations between common single nucleotide polymorphisms and complex traits and diseases [reviewed in 106]. However, it is important to note that the variants identified to date only explain a small portion of the genetic risk. This is due to several factors: (1) the limited power of GWA studies to identify common variants with very small effects (e.g. those increasing disease risk by a factor of 1.1 or less) [106]; (2) the limited ability of previous commercial microarray genotyping platforms to identify structural variation that may increase disease risk [107], and (3) the lack of power of GWA studies to identify rare casual variants [108]. However, our understanding of the genetic architecture of complex traits and diseases will dramatically improve in the near future. The application of GWA approaches to increasingly larger cohorts will make it possible to identify common variants of small effect that have not been detected in previous genome-wide scans. The development of better maps of structural variation in the human genome and improved microarray technologies (such as hybrid microarrays containing single nucleotide polymorphism and CNV probes as well as higher resolution genome comparative hybridization arrays) will clarify the relative role of structural variation in complex traits. Finally, the impressive advances in next generation sequencing technologies [109] are already opening the door to large-scale re-sequencing studies, which will dramatically expand our understanding of the genetics of complex traits and diseases, including the role of rare variants.

The coming years will undoubtedly reveal many additional candidate genes and genomic mechanisms underlying human performance. However, individual genes and specific mechanisms will still probably only explain a small portion of the variance for a particular athletic trait. For example, in our own work with the ACE gene, we detected a highly significant association of the ACE I-allele with higher arterial saturation (SaO_2) during exercise at high altitude ($p < 0.01$) [110]. The ACE gene itself explained only about 4% of the total variance in SaO_2 compared with the effect of acclimatization (explaining nearly 44% of the variance). The remaining unexplained variance was attributable to measurement error, other uninvestigated genetic, genomic, and environmental effects, epistasis, and GE interaction. Again, the interaction effects may be substantial. For example, Hagberg et al. [111] detected a significant association of the ACE genotype with exercise systolic blood pressure, but the association was evident only in untrained versus trained women. In other words, the ACE gene effect in that study was a function of the training environment and it was detectable only because the authors took training into account in their study design.

Most environmental effects are poorly characterized and thus difficult to account for in genetic study designs, particularly those that act over the developmental period. Consider a study by Kajantie et al. [112] who measured oral glucose tolerance and ACE I/D genotype in 423 adult women. The I-allele was associated with higher insulin response, but this association was conditioned by birth weight, i.e. it was only significant in subjects of low birthweight (p for interaction = 0.003). Low birthweight was taken as a proxy for suboptimal intrauterine growth due to poor fetal nutrition. Thus, the study suggests that an early-life environmental experience (i.e. poor fetal nutrition) conditions the expression of genetic effects on the adult phenotype many years later. Conceptually, this is a clear example of GE interaction, but it is important cardiopulmonary systems [1]. The specific literature in support of this conceptual framework is reviewed in greater detail below.

Low Birthweight Predicts a Decrease in the Relative Lean Body Mass or Muscle Mass in Adulthood

Numerous studies show significant positive associations of birthweight with the absolute LBM, the fat-free mass (FFM), and/or the muscle mass, i.e. bigger babies grow up to be bigger adults. Therefore, the inference that FP causes a reduction in the relative amount of lean tissue requires adjustment for adult body size and/or a demonstration of commensurate increases in adiposity. Several studies of infants [140], children/adolescents [141, 142], and adults [132–134, 143–147] meet one or both of these criteria. For example the study of Singhal et al. [142] on 78 adolescent girls, which assessed body composition via bioelectrical impedance (BIA), showed that low birthweight was associated with lower FFM even after adjusting for adult height. Similarly, two BIA studies in adults showed that low birthweight predicts low FFM or LBM after adjusting for adult BMI or adult height [143, 147]. Two dual energy X-ray absorptiometry (DXA) studies in adults showed that low birthweight predicted low muscle mass, controlling for adult height [144], or that low birthweight predicted lower muscle mass commensurate with increased fat mass and percentage body fat [145]. Several other studies in this area are worth highlighting. The study by Kahn et al. [132] showed a strong association of birthweight with adult BMI. However, after adjusting for thigh muscle and bone area (assessed via anthropometrics) the strength of association dropped by 68%, as compared to a drop of only 30% when thigh fat area was used as a control variable. Thus, the authors concluded that the positive association of birthweight with BMI reflects increases in lean tissue as opposed to fat. In the study by Phillips [146] low birthweight was associated with smaller stature and body weight in adults, but much of the weight reduction was accounted for by a reduction in muscle mass as estimated by urinary creatinine excretion. Finally, studies by Loos et al. [133, 134] using a monozygotic twin approach showed that in young adult males and females, the heavier twin was taller and heavier as an adult, but when adjusted for body mass, heavier twins had greater LBM. Because monozygotic twins have the same genome and share certain aspects of the intrauterine environment, pair-wise twin comparisons allow for the control of genetic confounding factors and some potential maternal confounding factors. This strengthens the hypothesis that the amount of lean tissue in adulthood traces back to the twin-specific intrauterine experience.

Low Birthweight Predicts Increased Adiposity in Adulthood

The literature supporting increased adiposity in adults born with low birthweight is somewhat inconsistent. This may be due to the fact that the fat mass is more variable and subject to environmental effects in adulthood than the FFM. Also, increased adiposity is associated with both ends of the birth weight distribution, i.e. high birth weight is also associated with an increased prevalence of obesity in adults [148]. At least seven studies of adults and children using skin folds only to assess body composition show a negative association of birthweight with relative body fat [133, 134, 141, 149–152], but at least two others show no association or a weak association in one or more study groups [141, 153]. Despite the limitations of the skin fold technique, the method does have the advantage that fat patterning can be evaluated. Among the studies cited above where fat patterning was described, there is an apparent consensus that low birthweight is associated with increased truncal fat in adulthood as opposed to limb fat [133, 134, 141, 150–152]. A number of studies have employed more direct methods of body composition assessment. Using BIA, Eriksson et al. [143] showed that low birthweight was associated with an increased fat mass controlling for BMI in adults. However, Singhal et al. [142] failed to find a similar association in adolescent girls controlling for height. Using DXA, a case control study by Kensara et al. [145] showed that low-birthweight subjects (n = 32) had a significantly higher percent body fat than controls. However, two other DXA studies in much larger samples showed no association of low birthweight with increased body fat or body fat percentage [144, 154]. To date, the most methodologically comprehensive study in this area was that by Elia et al. [155] who measured body composition using a four-component model that included skin folds, DXA, air densitometry, and deuterium dilution to measure total body water in 85 children (6–9 years). Results were consistent across body composition measurement techniques and revealed birthweight to be a strong predictor of percentage body fat in prepubertal children, i.e. an increase in birthweight of 1 standard deviation was associated with a decrease of 1.95% body fat. Only one study, in adults, employed the 'gold-standard' technique of hydrodensitometry [154]. In that study, there was no association of birthweight with fat mass or percentage body fat in a sample of 231 nondiabetic Pima Indians. However, given the high prevalence of diabetes among adult Pima, the study sample may not have been representative of putative FP effects that take place in the general population.

Low Birthweight Predicts Decreased Bone Mass and/or Bone Mineral Density in Adulthood
There is also strong evidence that the development of bone is modified in utero, a topic that has recently been reviewed [135, 156]. For example, an epidemiological study by Gale et al. [144] in a cohort of 70- to 75-year-olds (n = 143) showed significant positive associations of birthweight with whole body bone mineral content after adjustment for age, sex, adult height, smoking, alcohol consumption, calcium intake, and physical activity. Similarly, the study of Labayen et al. [141] showed a significant positive association of birthweight with bone mass adjusting for height in adolescent girls, but not boys. Studies of infants suggest that these effects are apparent at birth. Lappillone et al. [140] showed significantly lower bone mineral content in small-for-gestational-age infants compared to appropriate-for-gestational-age controls. Also, Godfrey et al. [157] showed that newborn birthweight is positively associated with bone mass after adjusting for sex and gestational age.

Low Birthweight Predicts Decreased Muscle Strength in Adulthood
Birthweight is positively associated with grip strength, a measure that reflects the action/strength of forearm flexors. Because of the positive association of birthweight with muscle mass (see above), and because muscle force depends on muscle cross-sectional area, FP effects on muscle strength must be largely mediated by changes in LBM [158]. However, there is also some indication that FP affects muscle quality/muscle performance. Three large-scale epidemiological studies of middle-aged and elderly subjects by the same research group in the UK show strong positive associations of birthweight with grip strength even after adjustment for various adult body size covariates [135–137]. For example, the study by Kuh et al. [137] showed a positive association of birthweight with grip strength, after adjusting for age, sex, and adult height. The effect was present in both men and women, and a 1-kg difference in birthweight in the cohort meant going from the 10th to 80th percentile of the grip strength distribution. While these results certainly imply effects of FP on muscle function, no study to date has adjusted for a direct measure of muscle mass. Further, Maltin et al. [159] have reviewed evidence of FP effects on muscle fiber number, fiber type, and fiber size. Apart from severe manipulations such as maternal starvation [160, 161], there is little evidence that FP has effects on muscle histology, including fiber type proportions. Thus, more work is necessary in this area.

Does Low Birthweight Predict Other Performance Phenotypes?
Taylor et al. [162] showed an association of impaired fetal growth (low ponderal index) with muscle performance in adult women as assessed by ^{31}P magnetic resonance spectroscopy. Low ponderal index at birth was associated with biochemical differences in adults that were consistent with delayed activation of glycolysis/glycogenolysis at the commencement of hard exercise stressing the anaerobic system. Similarly, a near-infrared spectroscopy study by Thompson et al. [163] showed no association of ponderal index with muscle histology, but differences in muscle reoxygenation rate. These authors suggested that the reoxygenation rate differences by ponderal index were consistent with the delay in the activation of anaerobic glycolytic metabolism demonstrated by Taylor et al. [162].

Does Low Birthweight Predict Physical Activity/Inactivity or Components of Energy Expenditure?
FP effects on body composition may be mediated through aspects of lifelong energy balance and/or physical activity/inactivity. Recent animal research suggests that FP operates directly on behavioral traits including the caloric intake and the voluntary physical activity. In rats, maternal undernutrition during pregnancy acts *directly* to produce offspring that are less physically active and also hyperphagic throughout postnatal life [138, 139]. This is intriguing because it suggests that associations between obesity, the metabolic syndrome, cardiovascular diseases, sedentary behavior, and overeating may have a common biological cause that traces back to intrauterine

experience. In the context of athletic performance, physical activity pattern is obviously directly associated with many performance measures, but serious consideration should also be given to the potential role of child physical activity pattern as a determinant of adult performance capacity.

Two human studies have examined FP effects on components of energy expenditure and/or energy metabolism [154, 164]. One study of 449 middle-aged men and women reported a higher resting pulse rate in low-birthweight subjects equivalent to about 2.7 beats/min (decline) per kg birthweight increase [164]. The authors suggested that this was due to elevated sympathetic nervous system activity that was established in utero. However, an equally plausible explanation is that low-birthweight subjects are less physically active, thus not aerobically fit, and therefore have lower stroke volume and a higher resting pulse rate. The other study in 272 Pima Indians assessed muscle sympathetic nervous system activity directly using microneurography and found no association with birthweight [154]. This study also measured the 24-hour energy expenditure, the sleeping metabolic rate (SMR), and the 24-hour respiratory quotient in a metabolic chamber. Of these measures, only the SMR was significantly associated with birthweight ($r = -0.13$), i.e. SMR was marginally higher than predicted for body size and composition (by ~2–3%) in the lowest versus highest birthweight subjects. The study did not measure resting energy expenditure or basal metabolic rate, and subjects were inactive during their 24-hour stay in the metabolic chamber. It should be emphasized that SMR is not a proxy for basal metabolic rate, and the association detected could reflect sleep disturbances in low-birthweight individuals. Thus, to date there is little direct information available on links between birthweight and components of energy expenditure in humans.

Conclusions

The genomic information revolution has made it possible to interrogate in new and powerful ways the genetic basis of human athletic performance. There is no doubt that this area of inquiry will become a major focus within the exercise sciences. However, it is also true that the genetic blueprint of an individual operates under the influence of innumerable environmental effects, and many of these effects operate at the earliest stages of the life cycle. In recent years, there has been an equally important revolution in developmental biology with the recognition that early-life environmental experience has lasting effects on the adult phenotype. This chapter has reviewed FP as an example of a developmental effect, but similar developmental effects may operate during infancy, childhood, and/or adolescence. Unfortunately, there is a dearth of information on these later time periods with direct relevance to adult exercise capacity. In any case, whenever developmental effects act, there is evidence that these effects interact with an individual's genetic background to produce irreversible change. If true, then accounting for developmental effects will be a key not only for detecting genetic effects in future research but also for understanding the full range of variation in human athletic ability.

References

1 Teran-Garcia M, Rankinen T, Bouchard C: Genes, exercise, growth, and the sedentary, obese child. J Appl Physiol 2008;105:988–1001.
2 Hills AP, King NA, Armstrong TP: The contribution of physical activity and sedentary behaviours to the growth and development of children and adolescents: implications for overweight and obesity. Sports Med 2007;37:533–545.
3 Leamy LJ, Pomp D, Lightfoot JT: An epistatic genetic basis for physical activity traits in mice. J Hered 2008;99:639–646.
4 Patel DR, Greydanus DE: Genes and athletes. Adolesc Med 2002;13:249–255.
5 Myburgh KH: What makes an endurance athlete world-class? Not simply a physiological conundrum. Comp Biochem Physiol A Mol Integr Physiol 2003;136:171–190.
6 Heck AL, Barroso CS, Callie ME, Bray MS: Gene-nutrition interaction in human performance and exercise response. Nutrition 2004;20:598–602.
7 Macarthur DG, North KN: Genes and human elite athletic performance. Hum Genet 2005;116:331–339.
8 Rupert JL: The search for genotypes that underlie human performance phenotypes. Comp Biochem Physiol A Mol Integr Physiol 2003;136:191–203.
9 Bouchard D, Malina RM, Perusse L: Genetics of Fitness and Physical Performance. Champaign, Human Kinetics, 1997.
10 Bouchard C, Malina RM: Genetics of physiological fitness and motor performance. Exerc Sport Sci Rev 1983;11:306–339.
11 Bouchard C, Daw EW, Rice T, et al: Familial resemblance for VO2max in the sedentary state: the HERITAGE family study. Med Sci Sports Exerc 1998;30:252–258.
12 Bouchard C, An P, Rice T, et al: Familial aggregation of VO(2max) response to exercise training: results from the HERITAGE Family Study. J Appl Physiol 1999;87:1003–1008.
13 Rodas G, Calvo M, Estruch A, et al: Heritability of running economy: a study made on twin brothers. Eur J Appl Physiol Occup Physiol 1998;77:511–516.
14 Calvo M, Rodas G, Vallejo M, et al: Heritability of explosive power and anaerobic capacity in humans. Eur J Appl Physiol 2002;86:218–225.
15 Gaskill SE, Rice T, Bouchard C, et al: Familial resemblance in ventilatory threshold: the HERITAGE Family Study. Med Sci Sports Exerc 2001;33:1832–1840.
16 Tiainen K, Sipila S, Alen M, et al: Heritability of maximal isometric muscle strength in older female twins. J Appl Physiol 2004;96:173–180.
17 Beunen G, Thomis M: Gene powered? Where to go from heritability (h2) in muscle strength and power? Exerc Sport Sci Rev 2004;32:148–154.
18 Arden NK, Spector TD: Genetic influences on muscle strength, lean body mass, and bone mineral density: a twin study. J Bone Miner Res 1997;12:2076–2081.
19 Missitzi J, Geladas N, Klissouras V: Heritability in neuromuscular coordination: implications for motor control strategies. Med Sci Sports Exerc 2004;36:233–240.
20 Nordstrom P, Lorentzon R: Influence of heredity and environment on bone density in adolescent boys: a parent-offspring study. Osteoporos Int 1999;10:271–277.
21 Rice T, Daw EW, Gagnon J, et al: Familial resemblance for body composition measures: the HERITAGE Family Study. Obes Res 1997;5:557–562.
22 Perusse L, Bouchard C: Role of genetic factors in childhood obesity and in susceptibility to dietary variations. Ann Med 1999;31(suppl 1):19–25.
23 Rice T, Perusse L, Bouchard C, Rao DC: Familial aggregation of body mass index and subcutaneous fat measures in the longitudinal Quebec family study. Genet Epidemiol 1999;16:316–334.
24 Hong Y, Rice T, Gagnon J, et al: Familial clustering of insulin and abdominal visceral fat: the HERITAGE Family Study. J Clin Endocrinol Metab 1998;83:4239–4245.
25 Katzmarzyk PT, Perusse L, Bouchard C: Genetics of abdominal visceral fat levels. Am J Human Biol 1999;11:225–235.
26 Katzmarzyk PT, Malina RM, Perusse L, et al: Familial resemblance in fatness and fat distribution. Am J Human Biol 2000;12:395–404.
27 Suwa M, Nakamura T, Katsuta S: Heredity of muscle fiber composition and correlated response of the synergistic muscle in rats. Am J Physiol 1996;271:R432–R436.

28 Barrey E, Valette JP, Jouglin M, Blouin C, Langlois B: Heritability of percentage of fast myosin heavy chains in skeletal muscles and relationship with performance. Equine Vet J Suppl 1999;30:289–292.

29 Bouchard C, Simoneau JA, Lortie G, Boulay MR, Marcotte M, Thibault MC: Genetic effects in human skeletal muscle fiber type distribution and enzyme activities. Can J Physiol Pharmacol 1986;64:1245–1251.

30 Komi PV, Karlsson J: Physical performance, skeletal muscle enzyme activities, and fibre types in monozygous and dizygous twins of both sexes. Acta Physiol Scand Suppl 1979;462:1–28.

31 Komi PV, Viitasalo JH, Havu M, Thorstensson A, Sjodin B, Karlsson J: Skeletal muscle fibres and muscle enzyme activities in monozygous and dizygous twins of both sexes. Acta Physiol Scand 1977;100:385–392.

32 Bielen EC, Fagard RH, Amery AK: Inheritance of acute cardiac changes during bicycle exercise: an echocardiographic study in twins. Med Sci Sports Exerc 1991;23:1254–1259.

33 Bielen EC, Fagard RH, Amery AK: Inheritance of blood pressure and haemodynamic phenotypes measured at rest and during supine dynamic exercise. J Hypertens 1991;9:655–663.

34 Bielen E, Fagard R, Amery A: The inheritance of left ventricular structure and function assessed by imaging and Doppler echocardiography. Am Heart J 1991;121:1743–1749.

35 Bielen E, Fagard R, Amery A: Inheritance of heart structure and physical exercise capacity: a study of left ventricular structure and exercise capacity in 7-year-old twins. Eur Heart J 1990;11:7–16.

36 Rice T, Rao R, Perusse L, Bouchard C, Rao DC: Tracking of familial resemblance for resting blood pressure over time in the Quebec Family Study. Hum Biol 2000;72:415–431.

37 An P, Perusse L, Rankinen T, et al: Familial aggregation of exercise heart rate and blood pressure in response to 20 weeks of endurance training: the HERITAGE family study. Int J Sports Med 2003;24:57–62.

38 Beekman M, Heijmans BT, Martin NG, et al: Heritabilities of apolipoprotein and lipid levels in three countries. Twin Res 2002;5:87–97.

39 Poulsen P, Kyvik KO, Vaag A, Beck-Nielsen H: Heritability of type II (non-insulin-dependent) diabetes mellitus and abnormal glucose tolerance – a population-based twin study. Diabetologia 1999;42:139–145.

40 Poulsen P, Levin K, Petersen I, Christensen K, Beck-Nielsen H, Vaag A: Heritability of insulin secretion, peripheral and hepatic insulin action, and intracellular glucose partitioning in young and old danish twins. Diabetes 2005;54:275–283.

41 Freeman MS, Mansfield MW, Barrett JH, Grant PJ: Heritability of features of the insulin resistance syndrome in a community-based study of healthy families. Diabet Med 2002;19:994–999.

42 Goran MI: Genetic influences on human energy expenditure and substrate utilization. Behav Genet 1997;27:389–399.

43 Bouchard C, Tremblay A, Nadeau A, et al: Genetic effect in resting and exercise metabolic rates. Metabolism 1989;38:364–370.

44 Rice T, Tremblay A, Deriaz O, Perusse L, Rao DC, Bouchard C: Genetic pleiotropy for resting metabolic rate with fat-free mass and fat mass: the Quebec Family Study. Obes Res 1996;4:125–131.

45 Kramer AA: Heritability estimates of thoracic skeletal dimensions for a high-altitude Peruvian population; in Eckhardt RB, Melton TW (eds): Population Studies on Human Adaptation and Evolution in the Peruvian Andes. University Park, Pennsylvania State University, 1992, pp 25–49.

46 Mueller WH, Chakraborty R, Barton SA, Rothhammer F, Schull WJ: Genes and epidemiology in anthropological adaptation studies: familial correlations in lung function in populations residing different altitude in Chile. Med Anthropol 1980;4:367–384.

47 Beall CM, Blangero J, Williams-Blangero S, Goldstein MC: Major gene for percent of oxygen saturation of arterial hemoglobin in Tibetan highlanders. Am J Phys Anthropol 1994;95:271–276.

48 Beall CM, Strohl KP, Blangero J, et al: Quantitative genetic analysis of arterial oxygen saturation in Tibetan highlanders. Hum Biol 1997;69:597–604.

49 Beall CM, Almasy LA, Blangero J, et al: Percent of oxygen saturation of arterial hemoglobin among Bolivian Aymara at 3,900–4,000 m. Am J Phys Anthropol 1999;108:41–51.

50 An P, Rice T, Gagnon J, et al: A genetic study of sex hormone-binding globulin measured before and after a 20-week endurance exercise training program: the HERITAGE Family Study. Metabolism 2000;49:1014–1020.

51 Hong Y, Gagnon J, Rice T, et al: Familial resemblance for free androgens and androgen glucuronides in sedentary black and white individuals: the HERITAGE Family Study. Health, Risk Factors, Exercise Training and Genetics. J Endocrinol 2001;170:485–492.

52 An P, Rice T, Gagnon J, et al: Race differences in the pattern of familial aggregation for dehydroepiandrosterone sulfate and its responsiveness to training in the HERITAGE Family Study. Metabolism 2001;50:916–920.

53 Gusella JF, Wexler NS, Conneally PM, et al: A polymorphic DNA marker genetically linked to Huntington's disease. Nature 1983;306:234–238.

54 Risch N, Merikangas K: The future of genetic studies of complex human diseases. Science 1996;273: 1516–1517.
55 Bouchard C, Rankinen T, Chagnon YC, et al: Genomic scan for maximal oxygen uptake and its response to training in the HERITAGE Family Study. J Appl Physiol 2000;88:551–559.
56 Rankinen T, Perusse L, Borecki I, et al: The Na(+)-K(+)-ATPase alpha2 gene and trainability of cardiorespiratory endurance: the HERITAGE family study. J Appl Physiol 2000;88:346–351.
57 Rivera MA, Perusse L, Simoneau JA, et al: Linkage between a muscle-specific CK gene marker and VO2max in the HERITAGE Family Study. Med Sci Sports Exerc 1999;31:698–701.
58 Rankinen T, An P, Rice T, et al: Genomic scan for exercise blood pressure in the Health, Risk Factors, Exercise Training and Genetics (HERITAGE) Family Study. Hypertension 2001;38:30–37.
59 Rankinen T, An P, Perusse L, et al: Genome-wide linkage scan for exercise stroke volume and cardiac output in the HERITAGE Family Study. Physiol Genomics 2002;10:57–62.
60 Rice T, Rankinen T, Province MA, et al: Genome-wide linkage analysis of systolic and diastolic blood pressure: the Quebec Family Study. Circulation 2000;102:1956–1963.
61 Rankinen T, Rice T, Boudreau A, et al: Titin is a candidate gene for stroke volume response to endurance training: the HERITAGE Family Study. Physiol Genomics 2003;15:27–33.
62 Grunig E, Janssen B, Mereles D, et al: Abnormal pulmonary artery pressure response in asymptomatic carriers of primary pulmonary hypertension gene. Circulation 2000;102:1145–1150.
63 Chagnon YC, Rice T, Perusse L, et al: Genomic scan for genes affecting body composition before and after training in Caucasians from HERITAGE. J Appl Physiol 2001;90:1777–1787.
64 Rice T, Chagnon YC, Perusse L, et al: A genomewide linkage scan for abdominal subcutaneous and visceral fat in black and white families: The HERITAGE Family Study. Diabetes 2002;51:848–855.
65 Sun G, Gagnon J, Chagnon YC, et al: Association and linkage between an insulin-like growth factor-1 gene polymorphism and fat free mass in the HERITAGE Family Study. Int J Obes Relat Metab Disord 1999;23:929–935.
66 Lanouette CM, Chagnon YC, Rice T, et al: Uncoupling protein 3 gene is associated with body composition changes with training in HERITAGE study. J Appl Physiol 2002;92:1111–1118.
67 Lakka TA, Rankinen T, Weisnagel SJ, et al: A quantitative trait locus on 7q31 for the changes in plasma insulin in response to exercise training: the HERITAGE Family Study. Diabetes 2003;52:1583–1587.
68 Rankinen T, Perusse L, Rauramaa R, Rivera MA, Wolfarth B, Bouchard C: The human gene map for performance and health-related fitness phenotypes: the 2003 update. Med Sci Sports Exerc 2004;36:1451–1469.
69 Campbell H, Rudan I: Interpretation of genetic association studies in complex disease. Pharmacogenomics J 2002;2:349–360.
70 McCarthy MI, Abecasis GR, Cardon LR, et al: Genome-wide association studies for complex traits: consensus, uncertainty and challenges. Nat Rev Genet 2008;9:356–369.
71 Sladek R, Rocheleau G, Rung J, et al: A genome-wide association study identifies novel risk loci for type 2 diabetes. Nature 2007;445:881–885.
72 Saxena R, Voight BF, Lyssenko V, et al: Genome-wide association analysis identifies loci for type 2 diabetes and triglyceride levels. Science 2007;316: 1331–1336.
73 Scott LJ, Mohlke KL, Bonnycastle LL, et al: A genome-wide association study of type 2 diabetes in Finns detects multiple susceptibility variants. Science 2007;316:1341–1345.
74 Unoki H, Takahashi A, Kawaguchi T, et al: SNPs in KCNQ1 are associated with susceptibility to type 2 diabetes in East Asian and European populations. Nat Genet 2008.
75 Yasuda K, Miyake K, Horikawa Y, et al: Variants in KCNQ1 are associated with susceptibility to type 2 diabetes mellitus. Nat Genet 2008;40:1092–1097.
76 Easton DF, Pooley KA, Dunning AM, et al: Genome-wide association study identifies novel breast cancer susceptibility loci. Nature 2007;447:1087–1093.
77 Hunter DJ, Kraft P, Jacobs KB, et al: A genome-wide association study identifies alleles in FGFR2 associated with risk of sporadic postmenopausal breast cancer. Nat Genet 2007;39:870–874.
78 Stacey SN, Manolescu A, Sulem P, et al: Common variants on chromosome 5p12 confer susceptibility to estrogen receptor-positive breast cancer. Nat Genet 2008;40:703–706.
79 Stacey SN, Manolescu A, Sulem P, et al: Common variants on chromosomes 2q35 and 16q12 confer susceptibility to estrogen receptor-positive breast cancer. Nat Genet 2007;39:865–869.
80 Gold B, Kirchhoff T, Stefanov S, et al: Genome-wide association study provides evidence for a breast cancer risk locus at 6q22.33. Proc Natl Acad Sci USA 2008;105:4340–4345.

81 Eeles RA, Kote-Jarai Z, Giles GG, et al: Multiple newly identified loci associated with prostate cancer susceptibility. Nat Genet 2008;40:316–321.
82 Gudmundsson J, Sulem P, Rafnar T, et al: Common sequence variants on 2p15 and Xp11.22 confer susceptibility to prostate cancer. Nat Genet 2008;40:281–283.
83 Gudmundsson J, Sulem P, Manolescu A, et al: Genome-wide association study identifies a second prostate cancer susceptibility variant at 8q24. Nat Genet 2007;39:631–637.
84 Thomas G, Jacobs KB, Yeager M, et al: Multiple loci identified in a genome-wide association study of prostate cancer. Nat Genet 2008;40:310–315.
85 Yeager M, Orr N, Hayes RB, et al: Genome-wide association study of prostate cancer identifies a second risk locus at 8q24. Nat Genet 2007;39:645–649.
86 Choudhry S, Coyle NE, Tang H, et al: Population stratification confounds genetic association studies among Latinos. Hum Genet 2006;118:652–664.
87 Hoggart CJ, Parra EJ, Shriver MD, et al: Control of confounding of genetic associations in stratified populations. Am J Hum Genet 2003;72:1492–1504.
88 Parra EJ, Kittles RA, Shriver MD: Implications of correlations between skin color and genetic ancestry for biomedical research. Nat Genet 2004;36:S54–S60.
89 Pfaff CL, Kittles RA, Shriver MD: Adjusting for population structure in admixed populations. Genet Epidemiol 2002;22:196–201.
90 Iafrate AJ, Feuk L, Rivera MN, et al: Detection of large-scale variation in the human genome. Nat Genet 2004;36:949–951.
91 Sebat J, Lakshmi B, Troge J, et al: Large-scale copy number polymorphism in the human genome. Science 2004;305:525–528.
92 Tuzun E, Sharp AJ, Bailey JA, et al: Fine-scale structural variation of the human genome. Nat Genet 2005;37:727–732.
93 Redon R, Ishikawa S, Fitch KR, et al: Global variation in copy number in the human genome. Nature 2006;444:444–454.
94 Feuk L, Carson AR, Scherer SW: Structural variation in the human genome. Nat Rev Genet 2006;7:85–97.
95 Korbel JO, Urban AE, Affourtit JP, et al: Paired-end mapping reveals extensive structural variation in the human genome. Science 2007;318:420–426.
96 Hurles ME, Dermitzakis ET, Tyler-Smith C: The functional impact of structural variation in humans. Trends Genet 2008;24:238–245.
97 Lupski JR: Genomic rearrangements and sporadic disease. Nat Genet 2007;39:S43–S47.
98 Stranger BE, Forrest MS, Dunning M, et al: Relative impact of nucleotide and copy number variation on gene expression phenotypes. Science 2007;315:848–853.
99 Fellermann K, Stange DE, Schaeffeler E, et al: A chromosome 8 gene-cluster polymorphism with low human beta-defensin 2 gene copy number predisposes to Crohn disease of the colon. Am J Hum Genet 2006;79:439–448.
100 Yang Y, Chung EK, Wu YL, et al: Gene copy-number variation and associated polymorphisms of complement component C4 in human systemic lupus erythematosus (SLE): low copy number is a risk factor for and high copy number is a protective factor against SLE susceptibility in European Americans. Am J Hum Genet 2007;80:1037–1054.
101 Hollox EJ, Huffmeier U, Zeeuwen PL, et al: Psoriasis is associated with increased beta-defensin genomic copy number. Nat Genet 2008;40:23–25.
102 Weiss LA, Shen Y, Korn JM, et al: Association between microdeletion and microduplication at 16p11.2 and autism. N Engl J Med 2008;358:667–675.
103 Marshall CR, Noor A, Vincent JB, et al: Structural variation of chromosomes in autism spectrum disorder. Am J Hum Genet 2008;82:477–488.
104 Stefansson H, Rujescu D, Cichon S, et al: Large recurrent microdeletions associated with schizophrenia. Nature 2008;455:232–236.
105 Perry GH, Dominy NJ, Claw KG, et al: Diet and the evolution of human amylase gene copy number variation. Nat Genet 2007;39:1256–1260.
106 Altshuler D, Daly MJ, Lander ES: Genetic mapping in human disease. Science 2008;322:881–888.
107 McCarroll SA, Altshuler DM: Copy-number variation and association studies of human disease. Nat Genet 2007;39:S37–S42.
108 Bodmer W, Bonilla C: Common and rare variants in multifactorial susceptibility to common diseases. Nat Genet 2008;40:695–701.
109 Shendure J, Ji H: Next-generation DNA sequencing. Nat Biotechnol 2008;26:1135–1145.
110 Bigham AW, Kiyamu M, Leon-Velarde F, et al: Angiotensin-converting enzyme genotype and arterial oxygen saturation at high altitude in Peruvian Quechua. High Alt Med Biol 2008;9:167–178.
111 Hagberg JM, McCole SD, Brown MD, et al: ACE insertion/deletion polymorphism and submaximal exercise hemodynamics in postmenopausal women. J Appl Physiol 2002;92:1083–1088.
112 Kajantie E, Rautanen A, Kere J, et al: The effects of the ACE gene insertion/deletion polymorphism on glucose tolerance and insulin secretion in elderly people are modified by birth weight. J Clin Endocrinol Metab 2004;89:5738–5741.
113 Barlow DP: Gametic imprinting in mammals. Science 1995;270:1610–1613.
114 Razin A, Cedar H: DNA methylation and genomic imprinting. Cell 1994;77:473–476.

115 Cooney CA, Dave AA, Wolff GL: Maternal methyl supplements in mice affect epigenetic variation and DNA methylation of offspring. J Nutr 2002;132:2393S–2400S.
116 Ahuja N, Issa JP: Aging, methylation and cancer. Histol Histopathol 2000;15:835–842.
117 Petronis A: Human morbid genetics revisited: relevance of epigenetics. Trends Genet 2001;17:142–146.
118 Carroll JL: Developmental plasticity in respiratory control. J Appl Physiol 2003;94:375–389.
119 Bavis RW: Developmental plasticity of the hypoxic ventilatory response after perinatal hyperoxia and hypoxia. Respir Physiol Neurobiol 2005;149:287–299.
120 Brutsaert T, Parra E, Shriver M, et al: Effects of birth place and individual admixture on lung volume and exercise phenotypes of Peruvian Quechua. Am J Phys Anthro 2004;123:390–398.
121 Hsia CC, Johnson RL Jr, McDonough P, et al: Residence at 3,800-m altitude for 5 mo in growing dogs enhances lung diffusing capacity for oxygen that persists at least 2.5 years. J Appl Physiol 2007;102:1448–1455.
122 Fall CH, Osmond C, Barker DJ, et al: Fetal and infant growth and cardiovascular risk factors in women. BMJ 1995;310:428–432.
123 Phillips DI, Barker DJ, Hales CN, Hirst S, Osmond C: Thinness at birth and insulin resistance in adult life. Diabetologia 1994;37:150–154.
124 Barker DJ, Martyn CN, Osmond C, Hales CN, Fall CH: Growth in utero and serum cholesterol concentrations in adult life. BMJ 1993;307:1524–1527.
125 Phipps K, Barker DJ, Hales CN, Fall CH, Osmond C, Clark PM: Fetal growth and impaired glucose tolerance in men and women. Diabetologia 1993;36:225–228.
126 Barker DJ, Hales CN, Fall CH, Osmond C, Phipps K, Clark PM: Type 2 (non-insulin-dependent) diabetes mellitus, hypertension and hyperlipidaemia (syndrome X): relation to reduced fetal growth. Diabetologia 1993;36:62–67.
127 Hales CN, Barker DJ, Clark PM, et al: Fetal and infant growth and impaired glucose tolerance at age 64. BMJ 1991;303:1019–1022.
128 Jackson AA: Nutrients, growth, and the development of programmed metabolic function. Adv Exp Med Biol 2000;478:41–55.
129 Cameron N, Demerath EW: Critical periods in human growth and their relationship to diseases of aging. Am J Phys Anthropol 2002;35(suppl):159–184.
130 Barker DJ, Winter PD, Osmond C, Margetts B, Simmonds SJ: Weight in infancy and death from ischaemic heart disease. Lancet 1989;2:577–580.
131 Demerath EW, Guo SS, Chumlea WC, Towne B, Roche AF, Siervogel RM: Comparison of percent body fat estimates using air displacement plethysmography and hydrodensitometry in adults and children. Int J Obes Relat Metab Disord 2002;26:389–397.
132 Kahn HS, Narayan KM, Williamson DF, Valdez R: Relation of birth weight to lean and fat thigh tissue in young men. Int J Obes Relat Metab Disord 2000;24:667–672.
133 Loos RJ, Beunen G, Fagard R, Derom C, Vlietinck R: Birth weight and body composition in young women: a prospective twin study. Am J Clin Nutr 2002;75:676–682.
134 Loos RJ, Beunen G, Fagard R, Derom C, Vlietinck R: Birth weight and body composition in young adult men – a prospective twin study. Int J Obes Relat Metab Disord 2001;25:1537–1545.
135 Sayer AA, Cooper C: Fetal programming of body composition and musculoskeletal development. Early Hum Dev 2005;81:735–744.
136 Sayer AA, Cooper C, Evans JR, et al: Are rates of ageing determined in utero? Age Ageing 1998;27:579–583.
137 Kuh D, Bassey J, Hardy R, Aihie Sayer A, Wadsworth M, Cooper C: Birth weight, childhood size, and muscle strength in adult life: evidence from a birth cohort study. Am J Epidemiol 2002;156:627–633.
138 Vickers MH, Breier BH, McCarthy D, Gluckman PD: Sedentary behavior during postnatal life is determined by the prenatal environment and exacerbated by postnatal hypercaloric nutrition. Am J Physiol Regul Integr Comp Physiol 2003;285:R271–R273.
139 Vickers MH, Breier BH, Cutfield WS, Hofman PL, Gluckman PD: Fetal origins of hyperphagia, obesity, and hypertension and postnatal amplification by hypercaloric nutrition. Am J Physiol Endocrinol Metab 2000;279:E83–E87.
140 Lapillonne A, Braillon P, Claris O, Chatelain PG, Delmas PD, Salle BL: Body composition in appropriate and in small for gestational age infants. Acta Paediatr 1997;86:196–200.
141 Labayen I, Moreno LA, Blay MG, et al: Early programming of body composition and fat distribution in adolescents. J Nutr 2006;136:147–152.
142 Singhal A, Wells J, Cole TJ, Fewtrell M, Lucas A: Programming of lean body mass: a link between birth weight, obesity, and cardiovascular disease? Am J Clin Nutr 2003;77:726–730.
143 Eriksson J, Forsen T, Tuomilehto J, Osmond C, Barker D: Size at birth, fat-free mass and resting metabolic rate in adult life. Horm Metab Res 2002;34:72–76.

144 Gale CR, Martyn CN, Kellingray S, Eastell R, Cooper C: Intrauterine programming of adult body composition. J Clin Endocrinol Metab 2001;86:267–272.

145 Kensara OA, Wootton SA, Phillips DI, Patel M, Jackson AA, Elia M: Fetal programming of body composition: relation between birth weight and body composition measured with dual-energy X-ray absorptiometry and anthropometric methods in older Englishmen. Am J Clin Nutr 2005;82:980–987.

146 Phillips DI: Relation of fetal growth to adult muscle mass and glucose tolerance. Diabet Med 1995;12:686–690.

147 Yliharsila H, Kajantie E, Osmond C, Forsen T, Barker DJ, Eriksson JG: Body mass index during childhood and adult body composition in men and women aged 56–70 y. Am J Clin Nutr 2008;87:1769–1775.

148 Sorensen HT, Sabroe S, Rothman KJ, Gillman M, Fischer P, Sorensen TI: Relation between weight and length at birth and body mass index in young adulthood: cohort study. BMJ 1997;315:1137.

149 Zive MM, McKay H, Frank-Spohrer GC, Broyles SL, Nelson JA, Nader PR: Infant-feeding practices and adiposity in 4-y-old Anglo- and Mexican-Americans. Am J Clin Nutr 1992;55:1104–1108.

150 Barker M, Robinson S, Osmond C, Barker DJ: Birth weight and body fat distribution in adolescent girls. Arch Dis Child 1997;77:381–383.

151 Te Velde SJ, Twisk JW, Van Mechelen W, Kemper HC: Birth weight, adult body composition, and subcutaneous fat distribution. Obes Res 2003;11:202–208

152 Malina RM, Katzmarzyk PT, Beunen G: Birth weight and its relationship to size attained and relative fat distribution at 7 to 12 years of age. Obes Res 1996;4:385–390.

153 Sayer AA, Syddall HE, Dennison EM, et al: Birth weight, weight at 1 y of age, and body composition in older men: findings from the Hertfordshire Cohort Study. Am J Clin Nutr 2004;80:199–203.

154 Weyer C, Pratley RE, Lindsay RS, Tataranni PA: Relationship between birth weight and body composition, energy metabolism, and sympathetic nervous system activity later in life. Obes Res 2000;8:559–565.

155 Elia M, Betts P, Jackson DM, Mulligan J: Fetal programming of body dimensions and percentage body fat measured in prepubertal children with a 4-component model of body composition, dual-energy X-ray absorptiometry, deuterium dilution, densitometry, and skinfold thicknesses. Am J Clin Nutr 2007;86:618–624.

156 Davies JH, Evans BA, Gregory JW: Bone mass acquisition in healthy children. Arch Dis Child 2005;90:373–378.

157 Godfrey K, Walker-Bone K, Robinson S, et al: Neonatal bone mass: influence of parental birthweight, maternal smoking, body composition, and activity during pregnancy. J Bone Miner Res 2001;16:1694–1703.

158 Yliharsila H, Kajantie E, Osmond C, Forsen T, Barker DJ, Eriksson JG: Birth size, adult body composition and muscle strength in later life. Int J Obes (Lond) 2007;31:1392–1399.

159 Maltin CA, Delday MI, Sinclair KD, Steven J, Sneddon AA: Impact of manipulations of myogenesis in utero on the performance of adult skeletal muscle. Reproduction 2001;122:359–374.

160 Wilson SJ, Ross JJ, Harris AJ: A critical period for formation of secondary myotubes defined by prenatal undernourishment in rats. Development 1988;102:815–821.

161 Dwyer CM, Stickland NC: Does the anatomical location of a muscle affect the influence of undernutrition on muscle fibre number? J Anat 1992;181:373–376.

162 Taylor DJ, Thompson CH, Kemp GJ, et al: A relationship between impaired fetal growth and reduced muscle glycolysis revealed by 31P magnetic resonance spectroscopy. Diabetologia 1995;38:1205–1205.

163 Thompson CH, Sanderson AL, Sandeman D, et al: Fetal growth and insulin resistance in adult life: role of skeletal muscle morphology. Clin Sci (Lond) 1997;92:291–296.

164 Phillips DI, Barker DJ: Association between low birthweight and high resting pulse in adult life: is the sympathetic nervous system involved in programming the insulin resistance syndrome? Diabet Med 1997;14:673–677.

Tom Brutsaert, PhD
Department of Anthropology
The University at Albany, SUNY
1400 Washington Ave., Albany, NY 12222 (USA)
Tel. +1 518 442 7769, Fax +1 518 442 5710, E-Mail tbrutsae@csc.albany.edu

Genetics and Sports: An Overview of the Pre-Molecular Biology Era

M.W. Peeters[a,b] · M.A.I. Thomis[a] · G.P. Beunen[a] · R.M. Malina[c,d]

[a]Research Centre for Exercise and Health, Department of Biomedical Kinesiology, Faculty of Kinesiology and Rehabilitation Sciences, Katholieke Universiteit Leuven, Leuven, [b]Postdoctoral Fellow Research Foundation Flanders, Brussels, Belgium; [c]Department of Kinesiology and Health Education, University of Texas at Austin, Austin, Tex., and [d]Tarleton State University, Stephenville, Tex., USA

Abstract

Estimated genetic and environmental contributions to individual differences in physical performance phenotypes, responsiveness to intermittent, aerobic and strength training, and specific skill training protocols are the focus of this chapter. Data are derived primarily from twin and family studies, although methods of analysis vary considerably. Estimates of heritability span a wide range for several performance phenotypes and the responsiveness to training. This is explained, in part, by differences in sample characteristics and analytical strategies. Corresponding data for skill acquisition are very limited. Data dealing with the effects of age, sex, maturation and ethnicity on heritability estimates are lacking, and information on behavioral phenotypes that may be related to performance is not available.

Copyright © 2009 S. Karger AG, Basel

Performance phenotypes, such as muscular strength and power, aerobic performance and motor performance are essentially outcome variables based on tests and measurements conducted under standardized conditions. The process of performance is not ordinarily considered. Hence, it is difficult to make inferences about sport performances which are carried out under dynamic conditions far removed from those under which performance phenotypes are measured. Allowing for this caveat, this chapter considers genetic contributions to individual differences in physical performance. Research methodology and estimates of genetic and environmental contributions to variation in performance-related phenotypes are briefly considered. Genetic and environmental contributions to several performance phenotypes are then evaluated followed by a brief consideration of responsiveness of several performance phenotypes to training and genetic considerations in skill acquisition. Finally, the pre-molecular biology methodology is placed within the past, present and future contexts for the study of genetic epidemiology in human performance.

Methodological Considerations

Physical performances and related factors are evaluated on a continuous scale of measurement. Distributions in the general population are Gaussian or skewed, which is typical for quantitative, multifactorial phenotypes that are influenced by multiple genes (polygenic) and environmental factors. As such, the issue is not a question of 'nature' versus 'nurture' [1]. Both nature and nurture and their interactions are important to understanding performance phenotypes.

Two basic approaches have been used to study the genetic basis of performance phenotypes and related characteristics: the unmeasured genotype approach (top down) – the focus of this chapter, and the measured genotype approach (bottom up) – the focus of other chapters [1, 2]. When the measured genotype is not available, inferences about genetic influences on a phenotype are based on statistical analyses of the distributions of measures in related individuals and families based on the theoretical framework of biometrical genetics [3]. Two major strategies are used: twin studies and family studies. The former includes both monozygotic (MZ) and dizygotic (DZ) twins. Since MZ twins have an identical genetic background, they will be more similar (higher intra-pair correlation) in a trait that is under genetic control than DZ twins who share on average one-half of their genes. With twin data, genetic and environmental factors unique to the individual and shared within families can be estimated and under certain assumptions dominant genetic effects can also be identified. In family studies, the similarities among parents and offspring and among siblings are studied, although generational differences imply both age-specific genetic and environmental effects, which increase complexity of analysis. The family approach permits the identification of genetic plus cultural transmission of traits and estimates of maximum heritability. If data from more extended or combined pedigrees are available, more sophisticated models can be tested.

Assessment of heritability is based on the model that total variation (V_{tot}) in a phenotype is partitioned into genetic (V_G), common environmental (V_C) and individual-specific environmental (V_E) components ($V_{tot} = V_G + V_C + V_E$). Heritability ($h^2$) refers to the proportion of the total variation that can be attributed to genetic effects (V_G/V_{tot}). It is generally assumed that the effects of different genes are additive (a^2) meaning that the genotypic effect of the heterozygote genotype on the phenotype falls exactly between the genotypic effects of both homozygote genotypes. Dominance genetic effects refer to the interaction between alleles at the same locus (heterozygote effect does not fall exactly in the middle between the two homozygote genotype effects) and epistasis describes the interaction between alleles at different loci. The contribution of environmental factors shared by family members (common environmental factors, $c^2 = V_C/V_{tot}$) and the proportion of environmental factors that act on an individual can also be estimated ($e^2 = V_E/V_{tot}$).

Several assumptions should be met when using the additive model: no interaction between gene action and environment (different genotypes all react equally to similar environmental factors), no gene × environment correlation (similar exposure of environments for different genotypes), no gene × gene interaction, and finally no assortative mating for the trait studied (one assumes people mate randomly for the phenotype in question). In all likelihood, influences on performance phenotypes do not follow all assumptions, and gene action/environmental influences are more or less important at different ages, in each sex, in specific ethnic populations and/or in affluent versus developing countries. Longitudinal (transmission) models are needed to study age-specific influences on the decomposition of inter-age correlations within MZ and DZ pairs into genetically and environmentally transmitted or time-specific sources of variance [4–6].

The effects of gene × environment interaction in the responsiveness of individuals to physical training and specific skill training protocols is an important effect that is largely unstudied. Specific designs can be used to study this interaction. A set of MZ twins with similar baseline performance levels can be exposed to training regimen at a similar load and followed longitudinally. 'Repeated-measures' ANOVA can be used to quantify the variability of responses *between* different 'genetic pairs' versus the variability of training responses *within* the same 'genetic pair'. A significant F-ratio represents a significant genotype × training interaction, or genotype-dependent response to the training protocol [7]. Alternatively a longitudinal analysis of trained MZ and DZ twins can specifically test for the presence of 'training-specific' gene sets that are independent of gene sets that contribute to pre-training levels of performance [8]. A third alternative which permits the study of gene × environment interaction in twin studies without an intervention protocol is the inclusion of a measured environmental variable (e.g. hours of training per week) in the structural equation model [9]. This approach results in heritability estimates that vary with different levels of the measured environmental variable if gene × environment interaction is in fact present. The method requires large sample sizes and has not yet been applied with physical performance phenotypes.

Related Phenotypes

Body size, physique, proportions and composition are associated with performance in the general population and elite athletes. Among youth, individual differences in the timing and tempo of biological maturation are an additional source of variation [10, 11]. Although there have been considerable advances in behavioral genetics, behavioral phenotypes that may be associated with performance are not ordinarily considered in genetic analyses of performance phenotypes.

Evidence from family, sibling and twin studies indicates that a significant proportion of the variance in anthropometric characteristics (skeletal lengths, skeletal breadths, limb circumferences, bone mass) is genetically determined [2]. Heritabilities for stature from 19 samples of twins ranged from 0.69 to 0.96, with a mean and standard deviation of 0.85 ± 0.07. A corresponding estimate for segment lengths was similar, 0.84 ± 0.10, while those for skeletal breadths and limb circumferences were more variable and lower. The preceding estimates are based largely on data from well-nourished samples of European (White) ancestry, and the genetic contribution to the variance in adult stature tends to be higher in these populations [11].

Body mass is under less genetic control than stature, but when corrected for stature the genetic determination of mass is about 0.40. Heritabilities of the BMI (weight/height^2) in MZ twins reared apart ranged between 0.40 and 0.70, while estimates based on adoption and large family studies were lower for percentage body fat and skinfolds [12]. Other studies, however, have reported higher genetic contributions to indicators of body fatness, for example 0.70–0.94 for the BMI, waist hip ratio, sum of skinfolds and 'bio-electrical impedance' – estimates of fat mass based on structural equation modeling in a large twin cohort [13]. Estimates from several Canadian studies indicated a maximal heritability of 0.40 for various indicators of body fat distribution [2], while estimates based on a ratio of trunk to extremity skinfolds in a longitudinal study of Belgian twins ranged from 0.78 to 0.88 [6]. Differences in reported heritabilities stem in large part from adjustments that are made or not made in the respective analyses. Estimated heritabilities tend to be lower when indicators of fatness were adjusted for size and when indicators of fat distribution were adjusted for size and fat mass.

Physique is most often assessed with the Heath-Carter anthropometric protocol for estimating somatotype [14]. Heritability estimates from twin studies, based mainly on adolescents, are generally >0.70 for the three somatotype components with few exceptions in late adolescent samples [15]. Estimates from family studies tend to cluster around 0.40–0.55 in adolescents and adults. In a large sample of young adult twin pairs, sex-specific heritabilities were 0.28, 0.86 and 0.76 for endomorphy, mesomorphy and ectomorphy, respectively, in males; corresponding heritabilities for the three components were, respectively, 0.32, 0.82 and 0.70 in females [16]. More than 70% of the total variation in each somatotype component was explained by sources of variance shared between the three components in adult men and women, reflecting the need to analyze the heritabilities of somatotype components with a multivariate approach [16]. The results also indicate that the three components of somatotype estimated with the Heath-Carter anthropometric protocol are to a large extent under the control of an identical set of genes and identical environmental factors. This also explains the high intercorrelations among components [4–6, 17]. The high interrelationships among somatotype components are a major shortcoming of the Heath-Carter anthropometric method. This is related, in part, to the lack of trunk measurements, especially in the derivation of mesomorphy, and use of the same measurements in deriving each component. Skinfolds and height are used to derive endomorphy; limb circumferences corrected for skinfold thickness, limb skeletal breadths and height are used to derive mesomorphy; and height and weight are used to derive ectomorphy.

Variability in measures of biological maturation in children and adolescents also has a significant genetic component. Major characteristics of the adolescent growth and questionable zygosity determination, heritabilities for the vertical jump ranged between 0.82 and 0.93 for twin studies and 0.22 and 0.68 for family studies [31]. More recent twin studies based on the vertical and standing long jump or maximal power output in the Wingate test indicate comparable heritabilities, though analytical strategies vary. Estimated heritabilities were 0.67 for the squat jump, 0.45 for the countermovement jump and 0.74 for maximal power developed in 5 s in a Wingate test in 16 young adult male twin pairs [32]. Among adolescent twins, an age-adjusted heritability of 0.75 was noted for the vertical jump in a sample of 54 male and female twin pairs, though possible sex differences were not considered [33], while heritabilities of 0.97 and 0.85 for peak power on a Wingate test were noted in 15 pairs of pre-adolescent and 15 pairs of adolescent female twins, respectively [26]. Two recent studies applied structural equation modeling to data for adolescents and included five zygosity groups, which permitted testing for sex differences. In a sample of 158 Japanese twin pairs 10–15 years of age, a heritability of 0.66 was obtained for the standing long jump under an AE-model with no evidence for sex differences, although age differences were not corrected [20]. In a sample of 105 male and female twin pairs of the same age, univariate cross-sectional and longitudinal models were used in genetic analyses of vertical jump [5, 34]. Sex differences were noted with the cross-sectional approach; girls showed generally higher heritabilities than boys at most measurement occasions, 0.74–0.91 compared to 0.47–0.85, respectively, with one estimate of 0.0 at 16 years [34]. With data aligned on age at PHV, sex differences were less apparent and reached significance only 3 years after PHV, 0.89 in girls and 0.61 in boys [5]. Limiting the analysis to twins at 10 years of age and including data for parents suggested some genetic dominance was included in the heritability of 0.65, which was slightly lower than a heritability of 0.72 when only twins were used in the analyses [31]. In summary, data for adolescence do not provide evidence for clear age or sex differences either, although a divergence in heritabilities is suggested towards later adolescence in Belgian twins [5]. Explosive strength data for adults are lacking. A heritability estimate of 0.32 was observed in a bivariate analysis of leg extensor power in elderly female twins; however, in a univariate analysis with an AE model which did not fit the data significantly worse than an ACE model, the heritability was 0.61 and thus more comparable to studies in adolescents [23]. Heritability appears to be somewhat higher for explosive compared with static strength, although this may be due to the fact that most evidence is derived from twin studies without inclusion of other relatives.

Motor Performance

Although jumping tasks are included in the preceding section, they are also included in studies of motor performance. Estimated heritabilities for tests of sprinting, jumping and throwing based on studies of twins and siblings of school age are variable, ranging from 0.14 to 0.91 with no indication of sex differences [2, 35]. Most estimates are based on traditional analytical methods in contrast to more recent modeling strategies. Heritabilities for dashes vary somewhat with distance but do not support the conclusion that heritability is higher for shorter distances. An early study of parent-offspring similarities compared the performances of 24 college age men with that of their fathers when the latter were of college age 34 years earlier [36]. Correlations of 0.86 for the running long jump and 0.59 for the 100-yard dash suggest reasonably similar performances in these tasks when fathers and sons were about the same age in young adulthood. The results could be due to both inherited factors and/or the modeling of activity and sport behavior from parent to child.

Little is known of the genetic contribution to movement patterns. Results of two studies suggest a significant genotypic contribution to variation in the kinematic structure of the dash in twins of both sexes aged 11–15 years [37] and 8–14 years [38, 39]. Intrapair differences were consistently smaller in MZ compared to DZ males than in MZ compared to DZ females, suggesting that the sprinting performance of girls may be more amenable to environmental influences [37]. In contrast to the dash, intrapair differences for the kinetic structure of a throw for distance and the crawl stroke in swimming were similar in MZ and DZ twins [38, 39], suggesting an important role of environmental influences.

Balance is a skill that requires a combination of gross and fine motor control in the maintenance of equilibrium in static and dynamic tasks. Heritability estimates for various balance tests in samples of twins 8–18 years range from 0.27 to 0.86 [2, 35]. Estimates for a ladder climb and stabilometer (dynamic) are lower than those for a rail balance (static).

Flexibility measures are often indicated in performance test batteries and are joint specific. Data for biological relatives are not extensive. Estimated heritability for lower back flexibility in a sample of male twins 11–15 years was 0.69, whereas estimates for a combined sample of male and female twins 12–17 years ranged from 0.84, 0.70 and 0.91 for flexibility of the trunk, hip and shoulder, respectively. Familial correlations for lower back flexibility in biological siblings and parent-offspring pairs are low to moderate, ranging from 0.26 to 0.43 [2].

Aerobic Performance

Aerobic performance tests are both maximal and submaximal. Maximal aerobic power (maximal oxygen uptake) is measured under standardized laboratory conditions on a cycle ergometer or a treadmill and maximal or peak oxygen uptake is expressed per unit body mass. Submaximal aerobic performance is usually measured as the power output, at a heart rate of 150 or 170 beats per minute and is expressed per kg body mass (W/kg). Correlations for maximal oxygen uptake, adjusted for body mass, vary between 0.60 and 0.91 in MZ twins and between 0.04 and 0.53 in DZ twins

Multivariate models are also able to test hypotheses about the 'common action' of genes and environmental factors, which increases power to find genes with pleiotropic action. If, however, covariation between traits is due mainly to environmental factors, it may guide researchers to search for trait-specific genes rather than to focus on a search for pleiotropic genes.

MZ twin designs are very useful to study genotype × environment interaction with the unmeasured genotype approach. They are also useful in studying epigenetic variation that influences differences in levels of gene expression and might therefore help explain observed differences between MZ scores (discordance in disease status, but probably also in performance status) mediated in part by the interaction effects of physical activity and practice/training. Application of the method for testing genotype × environment interactions proposed by Purcell [9] to physical performance phenotypes may yield heritability estimates that vary with levels of physical activity or training. This approach may clarify some of the discrepancies among studies and different populations in heritability estimates for performance phenotypes.

References

1 Baker J, Davids K: Nature, nurture and sport performance. Int J Sport Perform 2007;38:1–117.
2 Bouchard C, Malina RM, Pérusse L: Genetics of Fitness and Physical Performance. Champaign, Human Kinetics, 1997.
3 Neale MC, Cardon LR: Methodology for Genetic Studies of Twins and Families, ed 1. Dordrecht, Kluwer, 1992.
4 Peeters MW, Thomis MA, Maes HH, Beunen GP, Loos RJ, Claessens AL, Vlietinck R: Genetic and environmental determination of tracking in static strength during adolescence. J Appl Physiol 2005;99:1317–1326.
5 Peeters MW, Thomis MA, Maes HH, Loos RJ, Claessens AL, Vlietinck R, Beunen GP: Genetic and environmental causes of tracking in explosive strength during adolescence. Behav Genet 2005;35:551–563.
6 Peeters MW, Beunen GP, Maes HH, Loos RJ, Claessens AL, Vlietinck R, Thomis MA: Genetic and environmental determination of tracking in subcutaneous fat distribution during adolescence. Am J Clin Nutr 2007;86:652–660.
7 Bouchard C, Perusse L, Leblanc C: Using MZ twins in experimental research to test for the presence of a genotype-environment interaction effect. Acta Genet Med Gemellol (Roma) 1990;39:85–89.
8 Thomis MA, Beunen GP, Maes HH, Blimkie CJ, Van LM, Claessens AL, Marchal G, Willems E, Vlietinck RF: Strength training: importance of genetic factors. Med Sci Sports Exerc 1998;30:724–731.
9 Purcell S: Variance components models for gene-environment interaction in twin analysis. Twin Res 2002;5:554–571.
10 Beunen G, Malina RM: Growth and biological maturation: relevance to athletic performance; in Hebestreit H, Bar-Or O (eds): The Young Athlete. Malden, Blackwell, 2008, pp 3–17.
11 Malina RM, Bar-Or O, Bouchard C: Growth, Maturation, and Physical Activity, ed 2. Champaign, Human Kinetics, 2004.
12 Katzmarzyk P, Bouchard C: Genetic influences on human body composition; in Heymsfield SB, Lohman TG, Wang Z, Going SB (eds): Human Body Composition, ed 2. Champaign, Human Kinetics, 2005, pp 243–257.
13 Souren NY, Paulussen AD, Loos RJ, Gielen M, Beunen G, Fagard R, Derom C, Vlietinck R, Zeegers MP: Anthropometry, carbohydrate and lipid metabolism in the East Flanders Prospective Twin Survey: heritabilities. Diabetologia 2007;50:2107–2116.
14 Carter JEL, Heath BH: Somatotyping – Development and Applications. Cambridge, Cambridge University Press, 1990.
15 Peeters MW, Thomis MA, Claessens AL, Loos RJ, Maes HH, Lysens R, Vanden EB, Vlietinck R, Beunen G: Heritability of somatotype components from early adolescence into young adulthood: a multivariate analysis on a longitudinal twin study. Ann Hum Biol 2003;30:402–418.

16 Peeters MW, Thomis MA, Loos RJ, Derom CA, Fagard R, Claessens AL, Vlietinck RF, Beunen GP: Heritability of somatotype components: a multivariate analysis. Int J Obes (Lond) 2007;31:1295–1301.

17 Claessens A, Beunen G, Simons J, Swalus P, Ostyn M, Renson R, Van Gerven D: A modification of Sheldon's anthroposcopic somatotype technique. Anthropol Kozl 1980;24:45–54.

18 Beunen G, Thomis M, Maes HH, Loos R, Malina RM, Claessens AL, Vlietinck R: Genetic variance of adolescent growth in stature. Ann Hum Biol 2000;27:173–186.

19 Beunen G, Thomis M: Gene powered? Where to go from heritability (h^2) in muscle strength and power? Exerc Sport Sci Rev 2004;32:148–154.

20 Okuda E, Horii D, Kano T: Genetic and environmental effects on physical fitness and motor performance. Int J Sport Health Sci 2005;3:1–9.

21 De MG, Thomis MA, Windelinckx A, Van LM, Maes HH, Blimkie CJ, Claessens AL, Vlietinck R, Beunen G: Covariance of isometric and dynamic arm contractions: multivariate genetic analysis. Twin Res Hum Genet 2007;10:180–190.

22 Silventoinen K, Magnusson PK, Tynelius P, Kaprio J, Rasmussen F: Heritability of body size and muscle strength in young adulthood: a study of one million Swedish men. Genet Epidemiol 2008;32:341–349.

23 Tiainen K, Sipila S, Alen M, Heikkinen E, Kaprio J, Koskenvuo M, Tolvanen A, Pajala S, Rantanen T: Shared genetic and environmental effects on strength and power in older female twins. Med Sci Sports Exerc 2005;37:72–78.

24 Frederiksen H, Gaist D, Petersen HC, Hjelmborg J, McGue M, Vaupel JW, Christensen K: Hand grip strength: a phenotype suitable for identifying genetic variants affecting mid- and late-life physical functioning. Genet Epidemiol 2002;23:110–122.

25 Thomis MA, Beunen GP, Van LM, Maes HH, Blimkie CJ, Claessens AL, Marchal G, Willems E, Vlietinck RF: Inheritance of static and dynamic arm strength and some of its determinants. Acta Physiol Scand 1998;163:59–71.

26 Maridaki M: Heritability of neuromuscular performance and anaerobic power in preadolescent and adolescent girls. J Sports Med Phys Fitness 2006;46:540–547.

27 Huygens W, Thomis MA, Peeters MW, Vlietinck RF, Beunen GP: Determinants and upper-limit heritabilities of skeletal muscle mass and strength. Can J Appl Physiol 2004;29:186–200.

28 Zhai G, Stankovich J, Ding C, Scott F, Cicuttini F, Jones G: The genetic contribution to muscle strength, knee pain, cartilage volume, bone size, and radiographic osteoarthritis: a sibpair study. Arthritis Rheum 2004;50:805–810.

29 Ropponen A, Levalahti E, Videman T, Kaprio J, Battie MC: The role of genetics and environment in lifting force and isometric trunk extensor endurance. Phys Ther 2004;84:608–621.

30 Arden NK, Spector TD: Genetic influences on muscle strength, lean body mass, and bone mineral density: a twin study. J Bone Miner Res 1997;12:2076–2081.

31 Maes HH, Beunen GP, Vlietinck RF, Neale MC, Thomis M, Vanden EB, Lysens R, Simons J, Derom C, Derom R: Inheritance of physical fitness in 10-yr-old twins and their parents. Med Sci Sports Exerc 1996;28:1479–1491.

32 Calvo M, Rodas G, Vallejo M, Estruch A, Arcas A, Javierre C, Viscor G, Ventura JL: Heritability of explosive power and anaerobic capacity in humans. Eur J Appl Physiol 2002;86:218–225.

33 Chatterjee S, Das N: Physical and motor fitness in twins. Jpn J Physiol 1995;45:519–534.

34 Beunen G, Thomis M, Peeters M, Maes HH, Claessens AL, Vlietinck R: Genetics of strength and power characteristics in children and adolescents. Pediatr Exerc Sci 2003;15:128–138.

35 Malina RM, Bouchard C: Genetic considerations in physical fitness; in Drury TF (ed): Assessing Physical Fitness and Physical Activity in Population-Based Surveys. Washington, US Government Printing Office, 1989, pp 453–473.

36 Cratty BJ: A comparison of fathers and sons in physical ability. Res Q 1960;31:12–15.

37 Sklad M: Similarity of movement in twins. Wychowanie Fizyczne Sport 1972;16:119–141.

38 Goya T, Hoshikawa T, Matsui H: Longitudinal Study on Selected Sports Performance Related with the Physical Growth and Development of Twins. Perth, University of Western Australia, 1991, pp 139–141.

39 Goya T, Amano Y, Hoshikawa T, Matsui H: Longitudinal study on the variation and development of selected sports performance in twins: case study for one pair of female monozygous (MZ) and dizygous (DZ) twins. Sport Sci 1993;14:151–168.

40 Bielen EC, Fagard RH, Amery AK: Inheritance of acute cardiac changes during bicycle exercise: an echocardiographic study in twins. Med Sci Sports Exerc 1991;23:1254–1259.

41 Bouchard C, Daw EW, Rice T, Perusse L, Gagnon J, Province MA, Leon AS, Rao DC, Skinner JS, Wilmore JH: Familial resemblance for VO2max in the sedentary state: the HERITAGE family study. Med Sci Sports Exerc 1998;30:252–258.

42 Fagard R, Bielen E, Amery A: Heritability of aerobic power and anaerobic energy generation during exercise. J Appl Physiol 1991;70:357–362.

43 Gaskill SE, Rice T, Bouchard C, Gagnon J, Rao DC, Skinner JS, Wilmore JH, Leon AS: Familial resemblance in ventilatory threshold: the HERITAGE Family Study. Med Sci Sports Exerc 2001;33:1832–1840.

44 Perusse L, Leblanc C, Bouchard C: Inter-generation transmission of physical fitness in the Canadian population. Can J Sport Sci 1988;13:8–14.

45 Perusse L, Gagnon J, Province MA, Rao DC, Wilmore JH, Leon AS, Bouchard C, Skinner JS: Familial aggregation of submaximal aerobic performance in the HERITAGE Family study. Med Sci Sports Exerc 2001;33:597–604.

46 Thibault MC, Simoneau JA, Cote C, Boulay MR, Lagasse P, Marcotte M, Bouchard C: Inheritance of human muscle enzyme adaptation to isokinetic strength training. Hum Hered 1986;36:341–347.

47 Prud'homme D, Bouchard C, Leblanc C, Landry F, Fontaine E: Sensitivity of maximal aerobic power to training is genotype-dependent. Med Sci Sports Exerc 1984;16:489–493.

48 Bouchard C, Tremblay A, Despres JP, Theriault G, Nadeau A, Lupien PJ, Moorjani S, Prudhomme D, Fournier G: The response to exercise with constant energy intake in identical twins. Obes Res 1994;2:400–410.

49 Hamel P, Simoneau JA, Lortie G, Boulay MR, Bouchard C: Heredity and muscle adaptation to endurance training. Med Sci Sports Exerc 1986;18:690–696.

50 Bouchard C, An P, Rice T, Skinner JS, Wilmore JH, Gagnon J, Perusse L, Leon AS, Rao DC: Familial aggregation of VO(2max) response to exercise training: results from the HERITAGE Family Study. J Appl Physiol 1999;87:1003–1008.

51 Simoneau JA, Lortie G, Boulay MR, Marcotte M, Thibault MC, Bouchard C: Inheritance of human skeletal muscle and anaerobic capacity adaptation to high-intensity intermittent training. Int J Sports Med 1986;7:167–171.

52 Sklad M: The genetic determination of the rate of learning of motor skills. Stud Phys Anthropol 1975;1:3–19.

53 Fox PW, Hershberger SL, Bouchard TJ Jr: Genetic and environmental contributions to the acquisition of a motor skill. Nature 1996;384:356–358.

54 Bray MS, Hagberg JM, Perusse L, Rankinen T, Roth SM, Wolfarth B, Bouchard C: The human gene map for performance and health-related fitness phenotypes: the 2006–2007 update. Med Sci Sports Exerc 2009;41:35–73.

Maarten W. Peeters, PhD
Department of Biomedical Kinesiology
Faculty of Kinesiology and Rehabilitation Sciences, Katholieke Universiteit Leuven
Tervuursevest 101, BE–3001, Leuven (Belgium)
Tel. +32 16 329 085, Fax +32 16 329 197, E-Mail Maarten.Peeters@Faber.kuleuven.be

Genes, Athlete Status and Training – An Overview

Ildus I. Ahmetov[a,b] · Viktor A. Rogozkin[a]

[a]Sports Genetics Laboratory, St Petersburg Research Institute of Physical Culture, St. Petersburg, and [b]Laboratory of Muscle Performance, SRC RF Institute for Biomedical Problems of the Russian Academy of Sciences, Moscow, Russia

Abstract

Significant data confirming the influence of genes on human physical performance and elite athlete status have been accumulated in recent years. Research of gene variants that may explain differences in physical capabilities and training-induced effects between subjects is widely carried out. In this review, the findings of genetic studies investigating DNA polymorphisms and their association with elite athlete status and training responses are reported. A literature search revealed that at least 36 genetic markers (located within 20 autosomal genes, mitochondrial DNA and Y-chromosome) are linked to elite athlete status and 39 genetic markers (located within 19 genes and mitochondrial DNA) may explain, in part, an interindividual variability of physical performance characteristics in response to endurance/strength training. Although more replication studies are needed, the preliminary data suggest an opportunity to use some of these genetic markers in an individually tailored prescription of lifestyle/exercise for health and sports performance.

Copyright © 2009 S. Karger AG, Basel

It has long been recognized that the interindividual variability of physical performance traits and the ability to become an elite athlete have a strong genetic basis. The genetic loci that influence these phenotypes are now being sought. As the result, several family, twin, case-control and cross-sectional studies suggested an important role of genetics along with epigenetic and environmental factors in determination of individual differences in athletic performance and training responses [1].

A quarter of a century has passed since the first use of the '*Genetics of fitness and physical performance*' term by Claude Bouchard [2] and the first publication in this field [3]; however, the era of modern genetics of physical performance began in late 1990s, when the data from the Human Genome Project acquired currency. Specifically, in 1997 Montgomery et al. [4] reported greater left ventricular (LV) hypertrophy in military recruits carrying the angiotensin I converting enzyme *(ACE)* DD genotype synthesis, and the degradation of vasodilator kinins. A polymorphism in intron 16 of the human *ACE* gene has been identified in which the presence (insertion, I allele) rather than the absence (deletion, D allele) of a 287-bp Alu-sequence insertion fragment is associated with lower serum and tissue ACE activity. An excess of the I allele has been associated with some aspects of endurance performance, being identified in 34 elite British ≥5,000 m distance runners [15] and 25 elite mountaineers [5]. In addition, an excess of the I allele is present in elite Australian (n = 64) [11], Croatian (n = 40) [16] and Russian (n = 107) [17] rowers as well as Spanish elite athletes (25 cyclists, 20 long-distance runners, 15 handball players) [18]. *ACE* I allele is also over-represented among 100 fastest Ironman triathletes [19], successful marathon runners (scoring on places from 1st to 150th) [20], 35 outstanding Russian middle-distance athletes (24 swimmers, 7 track-and-field endurance athletes, 4 cross-country skiers) [21], 33 Italian Olympic endurance athletes (10 road cyclists, 7 track-and-field runners, 16 cross-country skiers) [22], 80 Turkish endurance and power/endurance athletes (17 middle-distance running, 10 basketball, 18 handball, 35 football players) [23] and 16 long-distance (25 km) swimmers from different nationalities [24].

An excess frequency of the *ACE* I allele or II genotype in endurance-oriented athletes may be partly explained by a genotype-dependent improvement in skeletal muscle mechanical efficiency with training [25], association of the *ACE* II genotype with an increased percentage of slow-twitch type I fibers in human skeletal muscle [26], higher VO_{2max} [27], higher aerobic work efficiency [28], improved fatigue resistance [5], higher peripheral tissue oxygenation during exercise [29], greater aerobic power response to training [30], improved hypoxic ventilatory response [31], greater cardiac output and maximal power output in athletes [17, 32].

ADRA2A 6.7-kb Allele
The α-2A-adrenergic receptor (ADRA2A) plays a central role in the regulation of systemic sympathetic activity and hence cardiovascular responses such as heart rate (HR) and blood pressure (BP). The restriction enzyme *Dra*I identifies a restriction fragment length polymorphism in the 3′-untranslated region (6.7-/6.3-kb polymorphism) of the *ADRA2A* gene. Wolfarth et al. [33] have observed a significant difference in genotype distributions between elite endurance athletes (148 Caucasian male subjects) and sedentary controls (149 unrelated sedentary male subjects). A higher frequency of the 6.7-kb allele was found in athletes compared with the sedentary controls group. It was concluded that genetic variation in the *ADRA2A* gene or a locus in close proximity may play a role in being able to sustain the endurance training regimen necessary to attain a high level of maximal aerobic power [33].

ADRB2 Arg16 Allele
$β_2$ adrenergic receptor (encoded by *ADRB2*) is a member of the G protein-coupled receptor superfamily, expressed in many cell types throughout the body and play a pivotal role in the regulation of the cardiac, pulmonary, vascular, endocrine and

central nervous system. The Arg16Gly SNP (rs1042713 G/A) of the *ADRB2* gene and its association with several phenotypes has been described. Specifically, Arg16 allele is associated with lower receptor density and resting cardiac output [34]. Recently, Wolfarth et al. [35] reported that Arg16 allele was overrepresented in 313 white male elite endurance athletes compared with 297 white male sedentary controls, suggesting a positive association between the tested Arg16Gly polymorphism and endurance performance. The results of this study are in agreement with the previous work in which the association of Arg16 allele with higher peak VO_2 in heart failure patients was reported [36].

AMPD1 C34 Allele

Adenosine monophosphate deaminase 1 (AMPD1) catalyzes the deamination of adenosine monophosphate to inosine monophosphate in skeletal muscle. Deficiency of the AMPD1 is apparently a common cause of exercise-induced myopathy and probably the most common cause of metabolic myopathy in the human. In the overwhelming majority of cases, AMPD1 deficiency is due to a 34C/T transition in exon 2 (rs17602729) of the *AMPD1* gene, which creates a non-sense codon (Q12X) that prematurely terminates translation. AMPD1 deficiency individuals exhibit a low AMP deaminase activity, a faster accumulation of blood lactate during the early recovery from a 30-second sprint exercise [37] and reduced submaximal aerobic capacity (VO_2 at the ventilatory threshold, VT) [38]. In a study of Rico-Sanz et al. [39], subjects with the *AMPD1* XX genotype had diminished exercise capacity and cardiorespiratory responses to exercise in the sedentary state. Furthermore, the training response of ventilatory phenotypes during maximal exercise was more limited in XX [39]. Additionally, Fischer et al. [40] revealed a faster power decrease in the AMPD-deficient group during the 30-second Wingate cycling test. Finally, Rubio et al. [12] reported low frequency of the mutant X allele in a group of top-level Caucasian (Spanish) male endurance athletes (cyclists and runners, n = 104) compared with Spanish healthy non-athletes (n = 100).

BDKRB2 –9 Allele

Bradykinin is a potent endothelium-dependent vasodilator and acts via the bradykinin B2 receptor (encoded by *BDKRB2*). The absence (–9), rather than the presence (+9), of a 9-bp repeat sequence in exon 1 has previously been shown to be associated with increased gene transcription and higher *BDKRB2* mRNA expression. Williams et al. [41] have shown that the –9 allele of the *BDKRB2* gene is associated with higher efficiency of muscular contraction (i.e. the energy used per unit of power output during exercise or delta efficiency). In the same study of 81 elite British runners, analysis revealed a linear trend of increasing –9 allele frequency with distance run. The proportion of –9 alleles increased from 0.382 to 0.412 to 0.569 for those athletes running ≤200, 400–3,000, and ≥5,000 m, respectively [41]. The –9/–9 genotype of the *BDKRB2* gene was also overrepresented in male Caucasian triathletes (n = 443) of the

2000 and 2001 South African Ironman Triathlons compared with male controls (n = 203) [42]. Additionally, when divided into tertiles according to their finishing times, the –9/–9 genotype was only overrepresented in the fastest tertile.

EPAS1 rs1867785 G and rs11689011 T Alleles
Endothelial PAS domain protein 1 (EPAS1) is a hypoxia-inducible transcription factor and plays an important role in the catecholamine and mitochondrial homeostasis, in the control of cardiac output and erythropoietin regulation. Recently, Henderson et al. [43] have examined the impact of DNA variants within *EPAS1* (also known as *HIF2A*) gene in a physiological model such as elite endurance athletes. The frequencies of the G (rs1867785 A/G) and T (rs11689011 C/T) alleles located within the large intron 1 of the *EPAS1* gene tended to be higher in short (event duration no less than 50 s), middle (from 50 s to 10 min) and long (from ~2 to 10 h) distance Australian endurance athletes [in order of increasing frequency: 127 Ironmen triathletes, 58 cyclists, 24 Olympic distance triathletes, 172 rowers, 42 swimmers (100–800 m) and 28 middle distance runners] in comparison with 444 controls. They have also identified three *EPAS1* haplotypes to be significantly associated with elite endurance athletes classified according to the power-time model of endurance. The presence of one (haplotype G: A-T-G-G) and the absence of another (haplotype F: G-C-C-G) at the same locus was observed in athletes involved in high-intensity maximal exercise of a duration between 50 s and 10 min. In addition, athletes involved in a sustained steady-state effort (from ~2 to 10 h) demonstrated the increased presence of a third (haplotype H: A-T-G-A) [43].

MtDNA Haplogroups
Patients with mutations in mtDNA commonly present with exercise intolerance, muscle weakness and increased production of lactic acid [44]. An association has been found between several mtDNA control region polymorphisms and endurance capacity in sedentary men [45], and between morph variants of MTND5 and the level of maximum oxygen uptake [46], suggesting that certain mtDNA lineages may contribute to good aerobic performance. In a study of Finnish elite endurance athletes (n = 52), an excess of mtDNA haplogroup H and the absence of haplogroup K and subhaplogroup J2 compared with 1,060 controls and 89 sprinters was reported [44]. Haplogroup T was significantly less frequent among 95 Spanish elite endurance athletes in comparison with 250 healthy male population controls [47]. Recently, Scott et al. [48] have shown a greater proportion of L0 haplogroups and lower proportion of L3* haplogroups in 70 Kenyan elite endurance athletes compared with controls (Kenyan population, n = 85).

NFATC4 Gly160 Allele
Nuclear factor of activated T cell calcineurin-dependent 4 (NFATC4) is a transcription factor that regulates cardiac hypertrophy, muscle fiber composition, glucose and

lipid homeostasis, mitochondrial biogenesis and hippocampal neuronal signaling. *NFATC4* gene (also known as NFAT3) Gly160Ala polymorphism (rs2229309 G/C) was shown to be associated with indexes of cardiac hypertrophy [49]. Specifically, a lower mean of LV mass and wall thickness were observed in carriers of the *NFATC4* 160Ala allele. In a study of 1,580 Russian athletes, the frequency of the Gly160 allele of the *NFATC4* gene was higher in elite endurance-oriented athletes (n = 351) than in the control group [50]. Furthermore, *NFATC4* Gly allele was associated with high values of aerobic performance (VO_{2max} and AT in % of VO_{2max} values) both in male and female Russian rowers [51].

NOS3 Glu298, 164-bp and 4B Alleles

Endothelial nitric oxide synthase (NOS3) generates NO in blood vessels and is involved in regulating vascular function. The *NOS3* gene contains a number of most frequently studied polymorphisms, such as Glu298Asp (E298D or G894T or rs1799983) in exon 7, microsatellite $(CA)_n$ repeats in intron 13, and 27-bp repeats in intron 4 (4B/4A) variations. Evidence suggests that the *NOS3* 298Asp allele is associated with reduced ecNOS activity, reduced basal NO production and vascular disease in several populations. Saunders et al. [42] investigated *NOS3* Glu298Asp polymorphism (in combination with the *BDKRB2* polymorphism) in 443 male Caucasian Ironman triathletes and 203 healthy Caucasian male control subjects. There was a tendency of the *NOS3* Glu298 allele combined with a *BDKRB2* –9/–9 genotype to be overrepresented in the fastest finishing triathletes (n = 40, 28.6%) compared with the control subjects (n = 28, 17.3%; p = 0.028) [42]. In the Genathlete study, Wolfarth et al. [52] have examined the contribution of the three above-mentioned polymorphisms to discriminate 316 elite endurance athletes from 299 sedentary controls. The frequency of the most common 164-bp allele of the $(CA)_n$ repeat was significantly higher in endurance athletes in comparison with controls (p = 0.007). In a study of 168 Russian rowers [17], no difference was found between the athletes and controls for the 27-bp repeat polymorphism, although none of the highly elite rowers had the *NOS3* 4A/4A genotype which has been reported to be unfavorable for high-altitude adaptation (as well as *NOS3* Glu/Glu genotype) [53]. In addition, cross-sectional study in 27 Russian rowers revealed the association of *NOS3* 4B/4B genotype with higher aerobic capacity [17].

PPARA rs4253778 G Allele

Peroxisome proliferator-activated receptor-α (PPARα) is a transcription factor that regulates lipid, glucose and energy homeostasis. Endurance training increases the use of nonplasma fatty acids and may enhance skeletal muscle oxidative capacity by PPARα regulation of gene expression. Exercise-induced LV growth in healthy young men was strongly associated with the intron 7 G/C (rs4253778) polymorphism of the *PPARA* gene [54]. Individuals homozygous for the C allele had a 3-fold greater and heterozygotes had a 2-fold greater increase in LV mass than G allele homozygotes, comparison with controls. More specifically, Yang et al. [99] for the first time have shown that the frequency of the *ACTN3* XX genotype was reduced in Australian power athletes (6 vs. 20%) compared with controls, whereas none of the Olympians or female power athletes had an XX genotype. These findings have been supported by the independent replications in case-control studies of elite Finnish sprint athletes (frequency of the XX genotype: 0 vs. 9.2%) [44], elite Greek track and field athletes (frequency of the RR genotype: 47.94 vs. 25.97%) [100], top-level professional soccer players, participating in the Spanish Championships (frequency of the RR genotype: 48.3 vs. 28.5%) [101], elite-level strength athletes from across the United States (frequency of the XX genotype: 6.7 vs. 16.3%) [102] and Russian power-oriented athletes (frequency of the XX genotype: 6.4 vs. 14.2%) [14]. The hypothesis that *ACTN3* 577R allele may confer some advantage in power performance events was supported by several cross-sectional studies in non-athletes including mouse models of the *ACTN3* deficiency [103–110]. Additionally, Vincent et al. [106] have shown that the percentage surface and number of type IIx (fast-twitch glycolytic) fibers was greater in the RR than the XX genotype group of young healthy men. Furthermore, it is supposed that the α-actinin-3 deficiency may also negatively influence the power component of competition performance at least in Russian endurance athletes [13].

HIF1A 582Ser Allele
Glycolysis is the central source of anaerobic energy in humans, and this metabolic pathway is regulated under low-oxygen conditions by the transcription factor hypoxia-inducible factor 1α (HIF1α; encoded by *HIF1A*). HIF1α controls the expression of several genes implicated in various cellular functions including glucose metabolism (glucose transporters and glycolytic enzymes). A missense polymorphism, Pro582Ser, is present in exon 12 (C/T at bp 85; rs11549465). The rare T allele is predicted to result in a proline to serine change in the amino acid sequence of the protein. This substitution increases HIF1α protein stability and transcriptional activity, and therefore, may improve glucose metabolism. Recently, Ahmetov et al. [111] investigated a hypothesis that *HIF1A* Pro582Ser genotype distribution may differ for controls and Russian sprint/strength athletes, for which anaerobic glycolysis is one of the most important sources of energy for power performance. The frequency of *HIF1A* 582Ser allele was significantly higher in weightlifters (n = 53) than in 920 controls (17.9 vs. 8.5%; p = 0.001) and increased with their levels of achievement [sub-elite (14.7%) → elite (18.8%) → highly elite (25.0%)]. Moreover, the 582Ser allele was significantly associated with an increased proportion of fast-twitch muscle fibers in m. vastus lateralis of all-round speed skaters [111].

PPARA rs4253778 C Allele
PPARα is a ligand-activated transcription factor that regulates the expression of genes involved in fatty acid uptake and oxidation, glucose and lipid metabolism, LV growth and control of body weight. Jamshidi et al. [54] have shown that British army recruits homozygous for the rare *PPARA* C allele of the rs4253778 (intron 7 G/C)

polymorphism had a 3-fold greater increase in LV mass in response to training than G allele homozygotes. The hypothesis that intron 7 C allele is associated with the hypertrophic effect due to influences on cardiac and skeletal muscle substrate utilization was supported by the findings that *PPARA* C allele is overrepresented in 180 Russian power-oriented athletes (27.2 vs. 16.4%, p = 0.0001, in comparison with 1,242 controls) and associated with an increased proportion of fast-twitch muscle fibers in m. vastus lateralis of 40 male controls [9].

PPARG 12Ala Allele

PPARγ (encoded by *PPARG*) plays a critical physiological role as a central transcriptional regulator of adipogenic and lipogenic programs, insulin sensitivity and glucose homeostasis. The 12Ala variant of the *PPARG* gene Pro12Ala polymorphism (rs1801282 C/G) is associated with decreased receptor activity [112], improved insulin sensitivity [112] and increased BMI in humans [55, 113]. The carriers of the 12Ala allele show better glycaemic response to exercise training [114], higher rates of skeletal muscle glucose uptake [115] and greater cross-sectional area of muscle fibers [116]. In a study of Russian power-oriented athletes (n = 260), a higher frequency (23.8 vs. 15.1%, p < 0.0001) of the *PPARG* 12Ala allele compared with 1,073 controls has been reported [116].

Combined Impact of Gene Variants on Elite Athlete Status

Human physical performance phenotypes are classical quantitative traits influenced by many gene variants and environmental factors. It is very important to note that each DNA locus can explain a very small proportion of the phenotypic variance (~0.1 to ~1%). Therefore, very large sample sizes are needed to detect associations and various combinatorial approaches should be used. Until now, few studies have sought to define or quantify the impact of multiple genotype combinations that influence human physical performance [17, 41, 42, 50, 117–119].

Williams et al. [41] have shown the evidence for interaction between *BDRRB2* –9/+9 and *ACE* I/D genotypes in 115 British subjects, with individuals who were carriers of *ACE* II + *BDRRB2* –9/–9 genotype combination having the highest efficiency of muscular contraction before training. Furthermore, the *ACE*(I)/*BDRRB2*(–9) ('high kinin receptor activity') haplotype was significantly associated with endurance event among elite British athletes (p = 0.003) [41]. Saunders et al. [42] found that the *NOS3* Glu298 allele combined with a *BDKRB2* –9/–9 genotype was overrepresented in the fastest finishing Ironman triathletes (28.6%) compared with the control subjects (17.3%; p = 0.028). Gómez-Gallego et al. [117] have shown that professional road cyclists with the most strength/power-oriented genotype combination (*ACE* DD + *ACTN3* RR/RX), had higher respiratory compensation threshold values than those with the intermediate combinations (II + RX/RR, p = 0.036; DD + XX, p = 0.0004) but similar to those with the II + XX genotype combination. In a study of 173 Russian rowers, the prevalent combination of *ACE* I/D, *ACTN3* R577X and *PPARA* intron 7 G/C genotypes in all groups was ID-RX-GG, and its frequency in elite rowers was different compared with controls (28.6 vs. 17.3%) [119]. Further, the total frequency of the *ACE* I, *ACTN3* R577, *UCP2* 55Val and *UCP3* rs1800849 T alleles in highly elite Russian rowers was 57.1% [p = 0.027 in comparison with controls (41.2%)] [17]. An increasing linear trend of the favorable for rower performance total allele frequency with increasing level of achievement has also been reported [41.9% (nonelite) → 43.0% (sub-elite) → 45.8% (elite) → 57.1% (highly elite)] [17]. In a study of 8 gene polymorphisms in 1,580 Russian athletes and 1,057 controls, it was shown that 66.7% of highly elite endurance-oriented athletes were carriers of 8–12 endurance-related alleles (*NFATC4* Gly160, *PPARGC1A* Gly482, *PPARGC1B* 203Pro, *PPP3R1* 5I, *TFAM* 12Thr, *VEGF* rs2010963 C, *UCP2* 55Val, *UCP3* rs1800849 T), while there were only 18.1% of such persons in the control group (p < 0.0001) [50].

Genes and Response to Training

Regular exercise brings about various effects on the human body including increases in endurance capacity and muscle strength/volume, and improvements in arterial stiffness. The effects from exercise differ greatly among individuals, depending on lifestyle factors and genetic backgrounds. An understanding of genetic backgrounds will help to clarify criteria of daily physical activities and appropriate exercise for individuals including athletes, making it possible to apply individualized preventive medicine and medical care. Moreover, health benefit will result from enhanced health, physical functioning and lifestyle improvement [120].

The process of exercise-induced adaptation in skeletal muscle involves a multitude of signaling mechanism initiating replication of specific DNA genetic sequences, enabling subsequent translation of the genetic message and ultimately generating a series of amino acids that form new proteins. The functional consequences of these adaptations are determined by training volume, intensity and frequency, and the half-life of the proteins [121]. It should be noted that at the molecular and cellular levels individuals with the same genotype respond more similarly to training than those with different genotypes, indicating that genes play an important role in determination of individual differences in response to training. The search for genetic markers of trainability status will likely be more productive than the investigation of molecular markers of the performance phenotype in the untrained state [122].

Candidate Genes for Response to Endurance Training

Prolonged endurance training elicits a variety of metabolic and morphological changes, including mitochondrial biogenesis, fast-to-slow fiber-type transformation and substrate metabolism. Endurance adaptation results in increased muscle glycogen stores and glycogen sparing at submaximal lactate kinetics and morphological alterations, including greater type I fiber proportions per muscle area, and increased capillary and mitochondrial density. Repeated bouts of endurance exercise result in altered expression of a multiplicity of gene products, resulting in an altered muscle phenotype with improved resistance to fatigue [123]. In addition, HR and BP recovery acceleration after exercise are also important cardiovascular adaptations to endurance training. For instance, postexercise HR recovery improves after endurance training in sedentary healthy subjects and patients with cardiovascular diseases; however, a substantial heritable component may also be involved in the regulation of HR behavior in response to training, and this may partly explain the large interindividual variation in HR recovery. Several studies reported the associations of gene polymorphisms with postexercise HR and BP recovery [reviewed in 1].

Research on aerobic endurance clearly shows that some individuals respond more to training than others [124]. In the same study, the maximal heritability estimate of the VO_{2max} response to training adjusted for age and sex was reported to be 47%. Genetically gifted athletes have a much greater response to training. Accordingly, evidence that many of the world's best endurance runners originate from distinct regions of Ethiopia and Kenya, rather than being evenly distributed throughout their respective countries appears to further sustain the idea that the success of East African runners is genetically mediated [125]. Studies have shown that African distance runners have reduced lactic acid accumulation in muscles, increased resistance to fatigue, and increased oxidative enzyme activity, which equates with high levels of aerobic energy production. It has been proposed that these geographical disparities in athlete production may reflect a genetic similarity among those populating these regions for an athletic genotype and phenotype [126]. There are at least 29 genetic markers (located within 12 autosomal genes and MtDNA) that have been shown to be associated with endurance phenotypes in response to training (table 2).

ACE is the most extensively studied gene in genetics of fitness and physical performance. The wide distribution and multifunctional properties of the angiotensin-converting enzyme in skeletal muscles and cardiovascular system suggest that ACE could be involved in various physiological conditions. Montgomery et al. [5] reported greater exercise performance improvements (duration of repetitive elbow flexion performance with a 15-kg barbell) in British army recruits with *ACE* II compared with DD genotypes after exercise training. Improvement was 11-fold greater for those of II than for those of DD genotype. In the following study of army recruits, Williams et al. [25] have shown that the presence of the *ACE* I allele may confer an enhanced mechanical efficiency in trained muscle (by the means of 11-week program of primarily aerobic physical training). Two studies reported the association of *ACE* genotypes with aerobic performance in a disease state. More specifically, Defoor et al. [30] have shown that aerobic power response to training was greater in patients with *ACE* II than in those with the D allele in coronary artery disease patients. In addition, the *ACE* I allele was associated with an

Table 2. Gene variants (genetic markers) for endurance/strength/anaerobic power trainability status

Gene	Location	Polymorphism	Allele (genotype/haplotype) associated with an increase in endurance/power phenotype
Gene variants for endurance trainability status			
ACE	17q23.3	Alu I/D	I
		T-3892C	T
AMPD1	1p13	Q12X (rs17602729 C/T)	Q12
APOE	19q13.2	Arg158Cys	158Cys (ApoE*2)
		Cys112Arg	112Arg (ApoE*4)
ATP1A2	1q21–q23	8.0/3.3 kb (α_2 exon 1)	8.0 kb
		10.5/4.3 kb (α_2 exon 21–22)	10.5 kb
CKM	19q13.2–q13.3	NcoI A/G	favorable: AG
			unfavorable: GG
GABPB1 (NRF2)	15q21.2	rs12594956 A/C	rs12594956 A
		rs8031031 C/T	rs8031031 T
		rs7181866 A/G	rs7181866 G
HBB	11p15.5	–551C/T	–551C
		+16C/G (rs10768683)	+16C
HIF1A	14q21-q24	Pro582Ser (rs11549465 C/T)	Pro582
MtDNA loci	MtDNA	variants at positions 16133, 12406, 13365, 13470, 15925, 16223, 16362	favorable: non-Cam at 16133, 16223, 16362
			unfavorable: non-Cam at 12406, 13365, 13470, 15925
NRF1	7q32	rs2402970 C/T	rs2402970 C
		rs6949152 A/G	rs6949152 A
PPARD	6p21.2-p21.1	rs2016520 T/C	rs2016520 T
		rs2267668 A/G	rs2267668 A
PPARGC1A	4p15.1	Gly482Ser (rs8192678 G/A)	Gly482
VEGFA	6p12	(C–2578A)/(G–1154A) / (G–634C) combinations	AAG and CGC haplotypes
Gene variants for strength/anaerobic power trainability status			
ACE	17q23.3	Alu I/D	D (dynamic strength), I (isometric strength)

Table 2. Continued

Gene	Location	Polymorphism	Allele (genotype/haplotype) associated with an increase in endurance/power phenotype
ACTN3	11q13.1	R577X (rs1815739 C/T)	R577
IGF1	12q22-q23	(CA)$_n$ (>192/192/<192)	192
IL15	4q31	rs1057972 A/T	rs1057972 T
IL15RA	10p15-p14	rs2296135 A/C	rs2296135 C
PPP3R1	2p15	promoter 5I/5D	5I/5I (in combination with IGF1 192 allele)
RETN	19p13.2	−420 C/G	−420 C
		398 C/T	398 C
		540 G/A	540 G
TNF	6p21.3	−308 G/A	−308 A

enhanced exercise capacity response to physical training in chronic obstructive pulmonary disease patients [127]. Cam et al. [128] in a study of 55 female nonelite Caucasian Turkish athletes, investigated the association between *ACE* genotype and short- and medium-duration aerobic endurance performance (HR reserve parameters and 30 min running speed) improvements in response to training regimen. It has been shown that the *ACE* II genotype was associated with better improvements in medium duration aerobic endurance performance whilst *ACE* DD genotype was supposed to be more advantageous in performance enhancement in shorter duration and higher intensity endurance activities. However, in contrast to the above-mentioned studies, Rankinen et al. [129] reported a greater VO_{2max} increase with training in the DD homozygotes in comparison with II homozygotes. Recently, Bae et al. [130] have shown the association of the TT genotype for the *ACE* T-3892C polymorphism with a greater VO_{2max} increase in response to 12-week endurance training in Korean women.

AMPD is an important regulator of muscle energy metabolism: by converting AMP into inosine monophosphate with liberation of ammonia, this enzyme displaces the equilibrium of the myokinase reaction towards ATP production. Rico-Sanz et al. [39] reported that subjects with the unfavorable XX genotype (nonsense Q12X (C34T) mutation) at the *AMPD1* gene had diminished exercise capacity and cardiorespiratory responses to exercise in the sedentary state. Furthermore, the 20 week endurance training response of ventilatory phenotypes during maximal exercise was more limited in such subjects.

Apolipoprotein E (ApoE) is essential for the normal catabolism of triglyceride-rich lipoprotein constituents and plays less defined roles in skeletal muscle and nervous tissue. The *APOE* gene is polymorphic with three major alleles, *APOE*2* (158Cys), *APOE*3* (Arg158 or Cys112), *APOE*4* (112Arg), which translate into three isoforms of the protein: normal – ApoE-ε3; dysfunctional – ApoE-ε2 and ApoE-ε4. Interestingly, Thompson et al. [131] have shown that after 6 months of exercise training VO_{2max} increased only by 5% in the *APOE* 3/3 subjects versus 13% in the 2/3 and 3/4 subjects (p < 0.01).

The Na^+-K^+-ATPase α_2 (encoded by *ATP1A2* gene) plays an important role in the maintenance of electrolyte balance in the working muscle and thus may contribute to endurance performance. Rankinen et al. [132] investigated the association between 8.0/3.3 kb (α_2 exon 1) and 10.5/4.3 kb (α_2 exon 21–22) polymorphisms of the *ATP1A2* gene and the response of VO_{2max} and maximal power output to 20 week of endurance training in 472 sedentary Caucasian subjects. Sibling-pair linkage analysis revealed associations of 8.0-kb and 10.5-kb alleles with greater increase of VO_{2max} in response to training.

The muscle-specific creatine kinase (CKM) enzyme is bound specifically to the M line of the myofibril subfragment, one of the heavy meromyosin in the vicinity of the myosin ATPase and to the outer membrane and vesicles of the sarcoplasmic reticulum, which might change the Ca^{2+} uptake and power of muscle [133]. Four studies aimed to investigate the association between *CKM* gene *Nco*I A/G polymorphism in the 3′ untranslated region and VO_{2max} responses to endurance training. Rivera et al. [134] reported a significantly lower ΔVO_{2max} to endurance training in both 160 Caucasian parents and 80 unrelated adult offspring with the rare GG genotype. In the following study of 277 full sib-pairs from 98 Caucasian nuclear families, the same group supported the hypothesis that the *CKM Nco*I A/G polymorphism may contribute to individual differences in the VO_{2max} response to endurance training [135]. However, Defoor et al. [136] have not observed the association between *Nco*I genotypes and aerobic power in 927 Caucasian coronary artery disease patients. On the other hand, Zhou et al. [133] reported greater improvement of running economy (RE) indexes (indicators of submaximal aerobic capacity) after 18-week 5,000 m running program in *CKM* AG heterozygotes than those with AA and GG genotypes.

The GA binding protein transcription factor, β-subunit 1 (GABPB1, according to the new nomenclature, August 2008, also known as NRF-2) protein is involved in activation of cytochrome oxidase expression and nuclear control of mitochondrial function. He et al. [137] have examined the association between the rs12594956 A/C, rs8031031 C/T and rs7181866 A/G polymorphisms in the *GABPB1* gene and endurance capacity at baseline and after an 18-week endurance training program in young Chinese men. The association between rs12594956 A, rs8031031 T, rs7181866 G alleles and higher training response in VO_{2max} at RE was reported. Moreover, those men carrying the ATG haplotype had 57.5% higher training response in VO_{2max} at RE than noncarriers.

Hemoglobin is the iron-containing oxygen transport metalloprotein in the red blood cells and has a crucial role in the determination of cardiorespiratory fitness. The α *(HBA)* and β *(HBB)* loci determine the structure of the two types of polypeptide chains in adult hemoglobin. He et al. [138] reported greater improvement of RE indexes after 18-week 5,000 m running program in *HBB* -551CC (−551C/T polymorphism) and +16CC (rs10768683 +16C/G polymorphism) subjects (military police recruits of Chinese Han origin) than those with other genotypes.

HIF1 is a transcription factor that acts to regulate the transcription of numerous genes involved in the processes of erythropoiesis, angiogenesis, and metabolism. HIF1 is composed of two subunits, one of which (HIF1α; encoded by *HIF1A*) is regulated by hypoxia. Prior et al. [139] have shown that *HIF1A* Pro/Pro homozygotes of the Pro582Ser polymorphism showed preservation of the ability to increase VO_{2max} through aerobic exercise training at each age (55, 60 and 65 years) level evaluated. Contrary to this, subjects carrying the Ser allele were able to increase VO_{2max} to a similar extent as Pro/Pro homozygotes at 55 years of age, but showed significantly less increase in VO_{2max} to aerobic exercise training than Pro/Pro homozygotes at 60 and 65 years of age.

Aerobic ATP generation by oxidative phosphorylation in the mitochondrial respiratory chain is a prerequisite for prolonged muscle exercise. Out of 480 subunits of the respiratory chain enzymes, 13 are encoded by mtDNA that is a 16,568-bp maternally inherited genome [44]. Dionne et al. [46] in a 12- to 20-week endurance training program study of 46 sedentary males, have shown that subjects with non-Cambridge (non-Cam) sequence variants at nucleotide position of 12406, 13365, 13470 or 15925 had lower trainability, whereas those with the non-Cam at 16133 had greater trainability. In another study of 55 sedentary males, Murakami et al. [45] reported a greater increase in VO_{2max} in response to 8-week endurance training program in subjects with non-Cam variants at 16223 and 16362.

The biogenesis of mitochondria requires the expression of a large number of genes, most of which reside in the nuclear genome. Nuclear respiratory factor 1 (encoded by *NRF1*, location: 7q32) acts on nuclear genes encoding respiratory subunits and components of the mitochondrial transcription and replication machinery [140]. Data from the HERITAGE Family Study showed suggestive evidence of linkage between baseline levels of maximum oxygen uptake (VO_{2max}) in healthy sedentary African-Americans and 7q32 [141]. He et al. [142] assessed the effect of 18 weeks of endurance training on the possible association between *NRF1* genotypes (rs2402970 C/T and rs6949152 A/G) and aerobic phenotypes (VO_{2max}, VT and metabolic cost of running at submaximum intensities, RE) in base-like state and in response to endurance training. They have found a greater trainability expressed as VT or RE in subjects with rs2402970 CC and rs6949152 AA genotypes.

PPARδ is a key regulator of fatty acid oxidation and energy uncoupling. Hautala et al. [61] in a study of black subjects reported a smaller endurance training-induced increase in aerobic capacity (VO_{2max} and maximal power output) in *PPARD* rs2016520

CC homozygotes compared with T allele carriers. Additionally, Stefan et al. [64] have shown the association of the *PPARD* rs2267668 A allele (A/G polymorphism) and a greater increase in individual anaerobic threshold after 9 months of aerobic training in sedentary subjects.

PPARγ coactivator 1α (encoded by *PPARGC1A*) has been established as an important regulator of mitochondrial content in skeletal muscle due to its apparent coactivation of multiple mitochondrial transcription factors. The rare 482Ser allele of the Gly482Ser SNP in *PPARGC1A* (rs8192678 G/A) has been reported to be associated with less increase in individual anaerobic threshold after 9 months of aerobic training in sedentary subjects [64].

VEGF is important for the maintenance of the skeletal muscle capillary supply, as well as the expansion of the capillary supply in response to exercise training. Prior et al. [87] tested the hypothesis that *VEGFA* promoter polymorphisms (C-2578A, G-1154A, G-634C) are related to *VEGFA* gene expression in human myoblasts and VO_{2max} before and after aerobic exercise training. Accordingly, they reported a higher *VEGFA* expression in human myoblasts with AAG and CGC haplotypes and a greater VO_{2max} in subjects with AAG and CGC haplotypes before and after aerobic exercise training.

Candidate Genes for Response to Resistance/Anaerobic Power Training

Adaptation to resistance training includes increased protein synthesis via regulatory changes in transcriptional and translational mechanisms, and in the production of muscle cells that are added to existing myofibers or combine and form new contractile filaments, each providing additional contractile machinery with which to generate force [143]. Significant intersubject variation is observed in the trainability of strength and anaerobic power characteristics. Evidence for a genotype-environment interaction was found for the increase in one repetition maximum (1RM), static strength, and concentric flexion at 120°/s after 10 weeks of high resistance strength training of arm flexors in male twins [144]. Studies also suggest that isometric and concentric muscle strength are under moderate to high genetic control with heritabilities ranging between 44 and 78% for isometric and 31 and 61% for concentric strength [145]. Several studies reported the association between individual differences in strength/anaerobic power phenotypes in response to resistance/anaerobic power training and gene variations (at least 10 genetic markers within 8 genes), though less frequently than the study on endurance performance (table 2).

Folland et al. [92] examined the effect of *ACE* genotype upon changes in strength of quadriceps muscles in response to 9 weeks of specific strength training of male subjects. They found a greater strength gains in subjects with the D allele. Giaccaglia et al. [93] studied muscle strength before and after an 18-month light weight lifting training and walking in older men and women in relation to *ACE* genotype. Following exercise training, individuals with the DD genotype showed greater gains in knee extensor strength compared with II individuals. Cam et al. [128] investigated the impact of the *ACE* I/D variant in a nonelite female cohort on the trainability of running speeds. It was concluded that the *ACE* DD genotype seems to be more advantageous in performance enhancement in shorter duration and higher intensity endurance activities. Pescatello et al. [94] studied the associations among *ACE* I/D polymorphism and the response to a 12-week unilateral, upper-arm resistance training program in the trained and untrained arms of 631 men and women. They reported greater improvements in maximal voluntary contraction (MVC; isometric strength) in the trained and untrained arms after training in carriers of the I allele. On the other hand, 1RM (dynamic strength) and muscle cross-sectional area increases were greater for *ACE* D allele carriers in response to untrained arm training.

Two studies examined the *ACTN3* R577X genotype in relation to a variety of strength-related measures in response to training. Clarkson et al. [103] studied associations between *ACTN3* genotype and muscle size and elbow flexor isometric and dynamic strength in a large group of men (n = 247) and women (n = 355) before and after a 12-week elbow flexor/extensor resistance training program. Women homozygous for the *ACTN3* X allele had lower baseline MVC compared with heterozygotes. Furthermore, women with the XX genotype demonstrated greater absolute and relative 1RM gains compared with the carriers of the RR genotype after resistance training. Delmonico et al. [104] investigated knee extensor concentric peak power (PP) before and after a 10-week unilateral knee extensor strength training intervention in older men and women. In men, absolute PP change with strength training in the RR homozygotes approached a significantly higher value than in the XX group. In women, relative PP change with strength training in the RR group was higher than in the XX group.

Insulin-like growth factor-I (encoded by *IGF1*) is a potent mitogen and anabolic agent important in the growth of various body tissues, including skeletal muscle. In a 10-week unilateral strength training study with 67 older Caucasian men and women, Kostek et al. [146] reported that carriers of the 192 allele of the *IGF1* CA repeat promoter polymorphism showed greater quadriceps-muscle strength gains compared with noncarriers. In a second study with 128 white and black men and women, Hand et al. [147] investigated the effect of *IGF1* CA repeat and *PPP3R1* I/D polymorphisms on the changes in strength and muscle volume in response to a 10-week single-leg knee extension strength training program. There was a significant combined gene effect, including both *IGF1* CA repeat main effect and *IGF1* CA repeat x *PPP3R1* I/D gene x gene interaction effect, on change in strength with training. Specifically, individuals homozygous for *PPP3R1* II who were also heterozygous for 192 allele for *IGF1* had significantly greater increases in strength and muscle volume with training than those homozygous for *PPP3R1* II, who were also noncarriers of the 192 allele for *IGF1*.

Interleukin-15 is one of the most abundant cytokines in skeletal muscle with anabolic properties. Pistilli et al. [148] examined associations between interleukin-15

7 De Moor MH, Spector TD, Cherkas LF, et al: Genome-wide linkage scan for athlete status in 700 British female DZ twin pairs. Twin Res Hum Genet 2007;10:812–820.
8 Brutsaert TD, Parra EJ: What makes a champion? Explaining variation in human athletic performance. Respir Physiol Neurobiol 2006;151:109–123.
9 Ahmetov II, Mozhayskaya IA, Flavell DM, et al: PPARα gene variation and physical performance in Russian athletes. Eur J Appl Physiol 2006;97:103–108.
10 Ahmetov II, Astratenkova IV, Rogozkin VA: Association of a PPARD polymorphism with human physical performance. Mol Biol 2007;41:776–780.
11 Gayagay G, Yu B, Hambly B, et al: Elite endurance athletes and the ACE I allele – the role of genes in athletic performance. Hum Genet 1998;103:48–50.
12 Rubio JC, Martin MA, Rabadan M, et al: Frequency of the C34T mutation of the AMPD1 gene in world-class endurance athletes: does this mutation impair performance? J Appl Physiol 2005;98:2108–2112.
13 Ahmetov II, Druzhevskaya AM, Astratenkova IV, et al: The ACTN3 R577X polymorphism in Russian endurance athletes. Br J Sports Med, DOI: 10.1136/bjsm.2008.051540.
14 Druzhevskaya AM, Ahmetov II, Astratenkova IV, Rogozkin VA: Association of the ACTN3 polymorphism with power athlete status in Russians. Eur J Appl Physiol 2008;103:631–634.
15 Myerson S, Hemingway H, Budget R, et al: Human angiotensin I-converting enzyme gene and endurance performance. J Appl Physiol 1999;87:1313–1316.
16 Jelakovic B, Kuzmanic D, Milicic D, et al: Influence of angiotensin converting enzyme (ACE) gene polymorphism and circadian blood pressure (BP) changes on left ventricle (LV) mass in competitive oarsmen. Am J Hypertens 2000;13:182A.
17 Ahmetov II, Popov DV, Astratenkova IV, et al: The use of molecular genetic methods for prognosis of aerobic and anaerobic performance in athletes. Hum Physiol 2008;34:338–342.
18 Alvarez R, Terrados N, Ortolano R, et al: Genetic variation in the renin-angiotensin system and athletic performance. Eur J Appl Physiol 2000;82:117–120.
19 Collins M, Xenophontos SL, Cariolou MA, et al: The ACE gene and endurance performance during the South African Ironman Triathlons. Med Sci Sports Exerc 2004;36:1314–1320.
20 Hruskovicová H, Dzurenková D, Selingerová M, et al: The angiotensin converting enzyme I/D polymorphism in long distance runners. J Sports Med Phys Fitness 2006;46:509–513.
21 Nazarov IB, Woods DR, Montgomery HE, et al: The angiotensin converting enzyme I/D polymorphism in Russian athletes. Eur J Hum Genet 2001;9:797–801.
22 Scanavini D, Bernardi F, Castoldi E, et al: Increased frequency of the homozygous II ACE genotype in Italian Olympic endurance athletes. Eur J Hum Genet 2002;10:576–577.
23 Turgut G, Turgut S, Genc O, et al: The angiotensin converting enzyme I/D polymorphism in Turkish athletes and sedentary controls. Acta Med (Hradec Kralove) 2004;47:133–136.
24 Tsianos G, Sanders J, Dhamrait S, et al: The ACE gene insertion/deletion polymorphism and elite endurance swimming. Eur J Appl Physiol 2004;92:360–362.
25 Williams AG, Rayson MP, Jubb M, et al: The ACE gene and muscle performance. Nature 2000;403:614.
26 Zhang B, Tanaka H, Shono N, et al: The I allele of the angiotensin-converting enzyme gene is associated with an increased percentage of slow-twitch type I fibers in human skeletal muscle. Clin Genet 2003;63:139–144.
27 Hagberg JM, Ferrell RE, McCole SD, et al: VO2 max is associated with ACE genotype in postmenopausal women. J Appl Physiol 1998;85:1842–1846.
28 Zhang X, Wang C, Dai H, et al: Association between angiotensin-converting enzyme gene polymorphisms and exercise performance in patients with chronic obstructive pulmonary disease. Respirology 2008;13:683–688.
29 Kanazawa H, Otsuka T, Hirata K, Yoshikawa J: Association between the angiotensin-converting enzyme gene polymorphisms and tissue oxygenation during exercise in patients with COPD. Chest 2002;121:697–701.
30 Defoor J, Vanhees L, Martens K, et al: The CAREGENE study: ACE gene I/D polymorphism and effect of physical training on aerobic power in coronary artery disease. Heart 2006;92:527–528.
31 Patel S, Woods DR, Macleod NJ, et al: Angiotensin-converting enzyme genotype and the ventilatory response to exertional hypoxia. Eur Respir J 2003;22:755–760.
32 Hagberg JM, McCole SD, Brown MD, et al: ACE insertion/deletion polymorphism and submaximal exercise hemodynamics in postmenopausal women. J Appl Physiol 2002;92:1083–1088.
33 Wolfarth B, Rivera MA, Oppert JM, et al: A polymorphism in the alpha2-adrenoceptor gene and endurance athlete status. Med Sci Sports Exerc 2000;32:1709–1712.

34 Snyder EM, Hulsebus ML, Turner ST, et al: Genotype related differences in beta2 adrenergic receptor density and cardiac function. Med Sci Sports Exerc 2006;38:882–886.

35 Wolfarth B, Rankinen T, Mühlbauer S, et al: Association between a beta2-adrenergic receptor polymorphism and elite endurance performance. Metabolism 2007;56:1649–1651.

36 Wagoner LE, Craft LL, Singh B, et al: Polymorphisms of the beta(2)-adrenergic receptor determine exercise capacity in patients with heart failure. Circ Res 2000;86:834–840.

37 Norman B, Sabina RL, Jansson E: Regulation of skeletal muscle ATP catabolism by *AMPD1* genotype during sprint exercise in asymptomatic subjects. J Appl Physiol 2001;91:258–264.

38 Rubio JC, Pérez M, Maté-Muñoz JL, et al: *AMPD1* genotypes and exercise capacity in McArdle patients. Int J Sports Med 2008;29:331–335.

39 Rico-Sanz J, Rankinen T, Joanisse DR, et al: HERITAGE Family Study: Associations between cardiorespiratory responses to exercise and the C34T *AMPD1* gene polymorphism in the HERITAGE Family Study. Physiol Genomics 2003;14:161–166.

40 Fischer H, Esbjörnsson M, Sabina RL, et al: AMP deaminase deficiency is associated with lower sprint cycling performance in healthy subjects. J Appl Physiol 2007;103:315–322.

41 Williams AG, Dhamrait SS, Wootton PTE, et al: Bradykinin receptor gene variant and human physical performance. J Appl Physiol 2004;96:938–942.

42 Saunders CJ, Xenophontos SL, Cariolou MA, et al: The bradykinin b2 receptor (*BDKRB2*) and endothelial nitric oxide synthase 3 (*NOS3*) genes and endurance performance during Ironman Triathlons. Hum Mol Genet 2006;15:979–987.

43 Henderson J, Withford-Cave JM, Duffy DL, et al: The *EPAS1* gene influences the aerobic-anaerobic contribution in elite endurance athletes. Hum Genet 2005;118:416–423.

44 Niemi AK, Majamaa K: Mitochondrial DNA and *ACTN3* genotypes in Finnish elite endurance and sprint athletes. Eur J Hum Genet 2005;13:965–969.

45 Murakami H, Ota A, Simojo H, et al: Polymorphisms in control region of mtDNA relates to individual differences in endurance capacity or trainability. Jpn J Physiol 2002;52:247–256.

46 Dionne FT, Turcotte L, Thibault MC, et al: Mitochondrial DNA sequence polymorphism, VO2max, and response to endurance training. Med Sci Sports Exerc 1991;23:177–185.

47 Castro MG, Terrados N, Reguero JR, et al: Mitochondrial haplogroup T is negatively associated with the status of elite endurance athlete. Mitochondrion 2007;7:354–357.

48 Scott RA, Fuku N, Onywera VO, et al: Mitochondrial haplogroups associated with elite Kenyan athlete status. Med Sci Sports Exerc 2009;41:123–128.

49 Poirier O, Nicaud V, McDonagh T, et al: Polymorphisms of genes of the cardiac calcineurin pathway and cardiac hypertrophy. Eur J Hum Genet 2003;11:659–664.

50 Ahmetov II, Hakimullina AM, Shikhova JV, Rogozkin VA: The ability to become an elite endurance athlete depends on the carriage of high number of endurance-related alleles. Eur J Hum Genet 2008;16(suppl 2):341.

51 Popov DV, Ahmetov II, Shikhova JV, et al: *NFATC4* gene polymorphism and aerobic performance in athletes. Eur J Hum Genet 2008;16(suppl 2):336.

52 Wolfarth B, Rankinen T, Mühlbauer S, et al: Endothelial nitric oxide synthase gene polymorphism and elite endurance athlete status: the Genathlete study. Scand J Med Sci Sports 2008; 18:485–490.

53 Ahsan A, Norboo T, Baig MA, et al: Simultaneous selection of the wild-type genotypes of the G894T and 4B/4A polymorphisms of *NOS3* associate with high-altitude adaptation. Ann Hum Genet 2005;69: 260–267.

54 Jamshidi Y, Montgomery HE, Hense HW, et al: Peroxisome proliferator-activated receptor α gene regulates left ventricular growth in response to exercise and hypertension. Circulation 2002;105:950–955.

55 Ahmetov II, Popov DV, Mozhayskaya IA, et al: Association of regulatory genes polymorphisms with aerobic and anaerobic performance of athletes. Ross Fiziol Zh Im I M Sechenova 2007;93:837–843.

56 Wang YX, Zhang CL, Yu RT, et al: Regulation of muscle fiber type and running endurance by PPARδ. PLoS Biol 2004;2:e294.

57 Skogsberg J, Kannisto K, Cassel TN, et al: Evidence that peroxisome proliferator-activated receptor delta influences cholesterol metabolism in men. Arterioscler Thromb Vasc Biol 2003;23:637–643.

58 Vanttinen M, Nuutila P, Kuulasmaa T, et al: Single nucleotide polymorphisms in the peroxisome proliferator-activated receptor delta gene are associated with skeletal muscle glucose uptake. Diabetes 2005; 54:3587–3591.

59 Aberle J, Hopfer I, Beil FU, Seedorf U: Association of peroxisome proliferator-activated receptor delta +294T/C with body mass index and interaction with peroxisome proliferator-activated receptor alpha L162V. Int J Obes (Lond) 2006;30:1709–1713.

60 Ahmetov II, Glotov AS, Lyubaeva EV, et al: Regulation of muscle fiber type composition by gene polymorphisms; in Hoppeler H, Reilly T, Tsolakidis E, Gfeller L, Klossner S (eds): Book Abs 11th Ann Cong ECSS, Lausanne, 2006, p 253.

61 Hautala AJ, Leon AS, Skinner JS, et al: Peroxisome proliferator-activated receptor-delta polymorphisms are associated with physical performance and plasma lipids: the HERITAGE Family Study. Am J Physiol Heart Circ Physiol 2007;292:2498–2505.

62 Ling C, Poulsen P, Carlsson E, et al: Multiple environmental and genetic factors influence skeletal muscle PGC-1alpha and PGC-1beta gene expression in twins. J Clin Invest 2004;114:1518–1526.

63 Ridderstråle M, Johansson LE, Rastam L, Lindblad U: Increased risk of obesity associated with the variant allele of the *PPARGC1A* Gly482Ser polymorphism in physically inactive elderly men. Diabetologia 2006;49:496–500.

64 Stefan N, Thamer C, Staiger H, et al: Genetic variations in *PPARD* and *PPARGC1A* determine mitochondrial function and change in aerobic physical fitness and insulin sensitivity during lifestyle intervention. J Clin Endocrinol Metab 2007;92:1827–1833.

65 Lucia A, Gomez-Gallego F, Barroso I, et al: *PPARGC1A* genotype (Gly482Ser) predicts exceptional endurance capacity in European men. J Appl Physiol 2005;99344–348.

66 Arany Z, Lebrasseur N, Morris C, et al: The transcriptional coactivator PGC-1β drives the formation of oxidative type IIX fibers in skeletal muscle. Cell Metab 2007;5:35–46.

67 Andersen G, Wegner L, Yanagisawa K, et al: Evidence of an association between genetic variation of the coactivator PGC-1beta and obesity. J Med Genet 2005;42:402–407.

68 Ling C, Wegner L, Andersen G, et al: Impact of the peroxisome proliferator activated receptor-gamma coactivator-1beta (PGC-1β) Ala203Pro polymorphism on in vivo metabolism, PGC-1β expression and fibre type composition in human skeletal muscle. Diabetologia 2007;50:1615–1620.

69 Wolfarth B, Fischer A, Döring F, et al: Zusammenhang zwischen Polymorphismen in den PPARgamma Co-Faktor-Genen und der Ausdauerleistungsfähigkeit. Dtsch Z Sportmed 2007;58:202.

70 Tang W, Arnett DK, Devereux RB, et al: Identification of a novel 5-base pair deletion in calcineurin B (*PPP3R1*) promoter region and its association with left ventricular hypertrophy. Am Heart J 2005;150:845–851.

71 Ahmetov II, Linde EV, Shikhova YuV, et al: The influence of calcineurin gene polymorphism on morphofunctional characteristics of cardiovascular system of athletes. Ross Fiziol Zh Im I M Sechenova 2008;94:915–922.

72 Chow LS, Greenlund LJ, Asmann YW, et al: Impact of endurance training on murine spontaneous activity, muscle mitochondrial DNA abundance, gene transcripts, and function. J Appl Physiol 2007;102:1078–1089.

73 Buemann B, Schierning B, Toubro S, et al: The association between the val/ala-55 polymorphism of the uncoupling protein 2 gene and exercise efficiency. Int J Obes Relat Metab Disord 2001;25:467–471.

74 Dalgaard LT, Pedersen O: Uncoupling proteins: functional characteristics and role in the pathogenesis of obesity and type II diabetes. Diabetologia 2001;44:946–965.

75 Astrup A, Toubro S, Dalgaard LT, et al: Impact of the v/v 55 polymorphism of the uncoupling protein 2 gene on 24-h energy expenditure and substrate oxidation. Int J Obes Relat Metab Disord 1999;23:1030–1034.

76 Clapham JC, Arch JR, Chapman H, et al: Mice overexpressing human uncoupling protein-3 in skeletal muscle are hyperphagic and lean. Nature 2000;406:415–418.

77 Pilegaard H, Ordway GA, Saltin B, Neufer PD: Transcriptional regulation of gene expression in human skeletal muscle during recovery from exercise. Am J Physiol Endocrinol Metab 2000;279:E806–E814.

78 Schrauwen P, Hesselink MK, Vaartjes I, et al: Effect of acute exercise on uncoupling protein 3 is a fat metabolism mediated effect. Am J Physiol Endocrinol Metab 2002;282:E11–E17.

79 Schrauwen P, Hesselink M: UCP2 and UCP3 in muscle controlling body metabolism. J Exp Biol 2002;205:2275–2285.

80 Halsall DJ, Luan J, Saker P, et al: Uncoupling protein 3 genetic variants in human obesity: the c-55t promoter polymorphism is negatively correlated with body mass index in a UK Caucasian population. Int J Obes Relat Metab Disord 2001;25:472–477.

81 Schrauwen P, Xia J, Walder K, et al: A novel polymorphism in the proximal *UCP3* promoter region: effect on skeletal muscle *UCP3* mRNA expression and obesity in male non-diabetic Pima Indians. Int J Obes Relat Metab Disord 1999;23:1242–1245.

82 Echegaray M, Rivera I, Wolfarth B, et al: Uncoupling protein 3 gene polymorphism and elite endurance athlete status: the Genathlete study. Med Sci Sports Exerc 2003;35(suppl 1):378.

83 Hudson DE, Mokone GG, Noakes TD, Collins M: The –55 C/T polymorphism within the *UCP3* gene and performance during the South African Ironman Triathlon. Int J Sports Med 2004;25:427–432.

84 Richardson RS, Wagner H, Mudaliar SR, et al: Human *VEGF* gene expression in skeletal muscle: effect of acute normoxic and hypoxic exercise. Am J Physiol 1999;277:2247–2252.

85 Gustafsson T, Puntschart A, Kaijser L, et al: Exercise-induced expression of angiogenesis-related transcription and growth factors in human skeletal muscle. Am J Physiol 1999;276:679–685.
86 Watson CJ, Webb NJ, Bottomley MJ, Brenchley PE: Identification of polymorphisms within the vascular endothelial growth factor (*VEGF*) gene: correlation with variation in VEGF protein production. Cytokine 2000;12:1232–1235.
87 Prior SJ, Hagberg JM, Paton CM, et al: DNA sequence variation in the promoter region of the *VEGF* gene impacts *VEGF* gene expression and maximal oxygen consumption. Am J Physiol Heart Circ Physiol 2006;290:1848–1855.
88 Ahmetov II, Khakimullina AM, Popov DV, et al: Polymorphism of the vascular endothelial growth factor gene (*VEGF*) and aerobic performance in athletes. Hum Physiol 2008;34:477–481.
89 Jobling MA, Tyler-Smith C: The human Y chromosome: an evolutionary marker comes of age. Nat Rev Genet 2003;4:598–612.
90 Moran CN, Scott RA, Adams SM, et al: Y chromosome haplogroups of elite Ethiopian endurance runners. Hum Genet 2004;115:492–497.
91 Williams AG, Day SH, Folland JP, et al: Circulating angiotensin converting enzyme activity is correlated with muscle strength. Med Sci Sports Exerc 2005; 37:944–948.
92 Folland J, Leach B, Little T, et al: Angiotensin-converting enzyme genotype affects the response of human skeletal muscle to functional overload. Exp Physiol 2000;85:575–579.
93 Giaccaglia V, Nicklas B, Kritchevsky S, et al: Interaction between angiotensin converting enzyme insertion/deletion genotype and exercise training on knee extensor strength in older individuals. Int J Sports Med 2008;29:40–44.
94 Pescatello LS, Kostek MA, Gordish-Dressman H, et al: *ACE* ID genotype and the muscle strength and size response to unilateral resistance training. Med Sci Sports Exerc 2006;38:1074–1081.
95 Hopkinson NS, Nickol AH, Payne J, et al: Angiotensin converting enzyme genotype and strength in chronic obstructive pulmonary disease. Am J Respir Crit Care Med 2004;170:395–399.
96 Charbonneau DE, Hanson ED, Ludlow AT, et al: *ACE* genotype and the muscle hypertrophic and strength responses to strength training. Med Sci Sports Exerc 2008;40:677–683.
97 Wagner H, Thaller S, Dahse R, Sust M: Biomechanical muscle properties and angiotensin-converting enzyme gene polymorphism: a model-based study. Eur J Appl Physiol 2006;98:507–515.
98 Woods D, Hickman M, Jamshidi Y, et al: Elite swimmers and the D allele of the *ACE* I/D polymorphism. Hum Genet 2001;108:230–232.

99 Yang N, MacArthur DG, Gulbin JP, et al: *ACTN3* genotype is associated with human elite athletic performance. Am J Hum Genet 2003;73:627–631.
100 Papadimitriou ID, Papadopoulos C, Kouvatsi A, Triantaphyllidis C: The *ACTN3* gene in elite Greek track and field athletes. Int J Sports Med 2007; 29:352–355.
101 Santiago C, González-Freire M, Serratosa L, et al: *ACTN3* genotype in professional soccer players. Br J Sports Med 2008;42:71–73.
102 Roth SM, Walsh S, Liu D, et al: The *ACTN3* R577X nonsense allele is under-represented in elite-level strength athletes. Eur J Hum Genet 2008;16:391–394.
103 Clarkson PM, Devaney JM, Gordish-Dressman H, et al: *ACTN3* genotype is associated with increases in muscle strength and response to resistance training in women. J Appl Physiol 2005;99:154–163.
104 Delmonico MJ, Kostek MC, Doldo NA, et al: Alpha-actinin-3 (*ACTN3*) R577X polymorphism influences knee extensor peak power response to strength training in older men and women. J Gerontol A Biol Sci Med Sci 2007;62:206–212.
105 Moran CN, Yang N, Bailey ME, et al: Association analysis of the *ACTN3* R577X polymorphism and complex quantitative body composition and performance phenotypes in adolescent Greeks. Eur J Hum Genet 2007;15:88–93.
106 Vincent B, De Bock K, Ramaekers M, et al: *ACTN3* (R577X) genotype is associated with fiber type distribution. Physiol Genomics 2007;32:58–63.
107 Walsh S, Liu D, Metter EJ, et al: *ACTN3* genotype is associated with muscle phenotypes in women across the adult age span. J Appl Physiol 2008;105:1486–1491.
108 Delmonico MJ, Zmuda JM, Taylor BC, et al: Association of the *ACTN3* genotype and physical functioning with age in older adults. J Gerontol A Biol Sci Med Sci 2008;63:1227–1234.
109 MacArthur DG, Seto JT, Chan S, et al: An Actn3 knockout mouse provides mechanistic insights into the association between α-actinin-3 deficiency and human athletic performance. Hum Mol Genet 2008;17:1076–1086.
110 Chan S, Seto JT, MacArthur DG, et al: A gene for speed: contractile properties of isolated whole EDL muscle from an α-actinin-3 knockout mouse. Am J Physiol Cell Physiol 2008;295:897–904.
111 Ahmetov II, Hakimullina AM, Lyubaeva EV, et al: Effect of *HIF1A* gene polymorphism on human muscle performance. Bull Exp Biol Med 2008;146: 351–353.
112 Deeb SS, Fajas L, Nemoto M, et al: Pro12Ala substitution in PPARγ2 associated with decreased receptor activity, lower body mass index and improved insulin sensitivity. Nat Genet 1998;20:284–287.

113 Masud S, Ye S: Effect of the peroxisome proliferator-activated receptor-γ gene Pro12Ala variant on body mass index: a meta-analysis. J Med Genet 2003;40:773–780.

114 Adamo KB, Sigal RJ, Williams K, et al: Influence of Pro12Ala peroxisome proliferator-activated receptor γ2 polymorphism on glucose response to exercise training in type 2 diabetes. Diabetologia 2005;48:1503–1509.

115 Vänttinen M, Nuutila P, Pihlajamäki J, et al: The effect of the Ala12 allele of the peroxisome proliferator-activated receptor-γ2 gene on skeletal muscle glucose uptake depends on obesity: a positron emission tomography study. J Clin Endocrinol Metab 2005;90:4249–4254.

116 Ahmetov II, Mozhayskaya IA, Lyubaeva EV, et al: *PPARG* gene polymorphism and human physical performance. Bull Exp Biol Med 2008;146:567–569.

117 Gómez-Gallego F, Santiago C, González-Freire M, et al: Endurance performance: genes or gene combinations? Int J Sports Med 2009;30:66–72.

118 Williams AG, Folland JP: Similarity of polygenic profiles limits the potential for elite human physical performance. J Physiol 2008;586:113–121.

119 Ahmetov II, Astratenkova IV, Druzhevskaya AM, Rogozkin VA: Combinatorial genetic analysis of physical performance in athletes. Eur J Hum Genet 2007;15(suppl 1):301.

120 Maeda S, Murakami H, Kuno S, et al: Individual variations in exercise training-induced physiological effects and genetic factors. Int J Sport Health Sci 2006;4:339–347.

121 Coffey VG, Hawley JA: The molecular bases of training adaptation. Sports Med 2007;37:737–763.

122 Bouchard C, Malina RM, Perusse L: Genetics of Fitness and Physical Performance. Champaign, Human Kinetics, 1997.

123 Adhihetty PJ, Irrcher I, Joseph AM, et al: Plasticity of skeletal muscle mitochondria in response to contractile activity. Exp Physiol 2003;88:99–107.

124 Bouchard C, An P, Rice T, et al: Familial aggregation of VO_{2max} response to exercise training: results from the HERITAGE Family Study. J Appl Physiol 1999;87:1003–1008.

125 Scott RA, Georgiades E, Wilson RH, et al: Demographic characteristics of elite Ethiopian endurance runners. Med Sci Sports Exerc 2003;35:1727–1732.

126 Scott RA, Pitsiladis YP: Genetics and the success of East African distance runners. Int Sport Med J 2006;7:172–186.

127 Gosker HR, Pennings HJ, Schols AM: *ACE* Gene Polymorphism in COPD. Am J Respir Crit Care Med 2004;170:572.

128 Cam S, Colakoglu M, Colakoglu S, et al: *ACE* I/D gene polymorphism and aerobic endurance development in response to training in a non-elite female cohort. J Sports Med Phys Fitness 2007;47:234–238.

129 Rankinen T, Pérusse L, Gagnon J, et al: Angiotensin-converting enzyme ID polymorphism and fitness phenotype in the HERITAGE Family Study. J Appl Physiol 2000;88:1029–1035.

130 Bae JS, Kang BY, Lee KO, Lee ST: Genetic variation in the renin-angiotensin system and response to endurance training. Med Princ Pract 2007;16:142–146.

131 Thompson PD, Tsongalis GJ, Seip RL, et al: Apolipoprotein E genotype and changes in serum lipids and maximal oxygen uptake with exercise training. Metabolism 2004;53:193–202.

132 Rankinen T, Pérusse L, Borecki I, et al: The Na(+)-K(+)-ATPase alpha2 gene and trainability of cardiorespiratory endurance: the HERITAGE family study. J Appl Physiol 2000;88:346–351.

133 Zhou DQ, Hu Y, Liu G, et al: Muscle-specific creatine kinase gene polymorphism and running economy responses to an 18-week 5000-m training programme. Br J Sports Med 2006;40:988–991.

134 Rivera MA, Dionne FT, Simoneau JA, et al: Muscle-specific creatine kinase gene polymorphism and VO2max in the HERITAGE Family Study. Med Sci Sports Exerc 1997;29:1311–1317.

135 Rivera MA, Pérusse L, Simoneau JA, et al: Linkage between a muscle-specific CK gene marker and VO2max in the HERITAGE Family Study. Med Sci Sports Exerc 1999;31:698–701.

136 Defoor J, Martens K, Matthijs G, et al: The Caregene study: muscle-specific creatine kinase gene and aerobic power in coronary artery disease. Eur J Cardiovasc Prev Rehabil 2005;12:415–417.

137 He Z, Hu Y, Feng L, Lu Y, Liu G, Xi Y, Wen L, McNaughton LR: NRF2 genotype improves endurance capacity in response to training. Int J Sports Med 2007;28:717–721.

138 He Z, Hu Y, Feng L, et al: Polymorphisms in the HBB gene relate to individual cardiorespiratory adaptation in response to endurance training. Br J Sports Med 2006;40:998–1002.

139 Prior SJ, Hagberg JM, Phares DA, et al: Sequence variation in hypoxia-inducible factor 1alpha (HIF1A): association with maximal oxygen consumption. Physiol Genomics 2003;15:20–26.

140 Kelly DP, Scarpulla RC: Transcriptional regulatory circuits controlling mitochondrial biogenesis and function. Genes Dev 2004;18:357–368.

141 Rico-Sanz J Rankinen T, Rice T et al: Quantitative trait for maximal exercise capacity phenotypes and their responses to training in the HERITAGE Family Study. Physiol Genom 2004;16:256–260.

142 He Z, Hu Y, Feng L et al: NRF-1 genotypes and endurance exercise capacity in young Chinese men. Br J Sports Med 2008;42:361–366.
143 Rennie MJ, Wackerhage H, Spangenburg EE, Booth FW: Control of the size of the human muscle mass. Annu Rev Physiol 2004;66:799–828.
144 Thomis MAI, Beunen GP, Maes HH, et al: Strength training: importance of genetic factors. Med Sci Sports Exerc 1998;30:724–731.
145 De Mars G, Windelinckx A, Huygens W, et al: Genome-wide linkage scan for contraction velocity characteristics of knee musculature in the Leuven Genes for Muscular Strength Study. Physiol Genomics 2008;35:36–44.
146 Kostek MC, Delmonico MJ, Reichel JB, et al: Muscle strength response to strength training is influenced by insulin-like growth factor 1 genotype in older adults. J Appl Physiol 2005;98:2147–2154.
147 Hand BD, Kostek MC, Ferrell RE, et al: Influence of promoter region variants of insulin-like growth factor pathway genes on the strength-training response of muscle phenotypes in older adults. J Appl Physiol 2007;103:1678–1687.
148 Pistilli EE, Devaney JM, Gordish-Dressman H, et al: Interleukin-15 and interleukin-15R alpha SNPs and associations with muscle, bone, and predictors of the metabolic syndrome. Cytokine 2008;43:45–53.
149 Pistilli EE, Gordish-Dressman H, Seip RL, et al: Resistin polymorphisms are associated with muscle, bone, and fat phenotypes in white men and women. Obesity 2007;15:392–402.
150 Nicklas BJ, Mychaleckyj J, Kritchevsky S, et al: Physical function and its response to exercise: associations with cytokine gene variation in older adults with knee osteoarthritis. Gerontol A Biol Sci Med Sci 2005;60:1292–1298.
151 Dennis C: Rugby team convert to give test a try. Nature 2005;434:260.

Ildus I. Ahmetov, PhD
Laboratory of Muscle Performance
SRC RF Institute for Biomedical Problems of the Russian Academy of Sciences
76A Khoroshevskoe chaussee, 123007 Moscow (Russia)
Tel. +791 6574 7544, E-Mail genoterra@mail.ru

Angiotensin-Converting Enzyme, Renin-Angiotensin System and Human Performance

David Woods

Newcastle and Northumbria NHS Trust, School of Clinical Medical Sciences, University of Newcastle upon Tyne, and Royal Army Medical Corps, Army Medical Services, Newcastle upon Tyne, UK

Abstract

The angiotensin-converting enzyme (ACE) I/D polymorphism is strongly associated with circulating ACE levels in European populations. Initial studies over 10 years ago suggested an association between the I-allele (associated with low circulating ACE activity) and elite endurance performance. Subsequent studies also then suggested an association of the D-allele (high circulating ACE) with power-oriented performance. Not all studies are concordant however. The published literature is beset with the problems attendant in interpreting population-association studies including case definition, adequate cohort size, selection bias and appropriate control groups. Notwithstanding this, a plethora of data has been published. This has been followed by various studies trying to elucidate a physiological mechanism for such an association. This chapter will review the available data regarding the ACE I/D polymorphism and human performance both at sea level and at high altitude. It will also evaluate any data pertaining to postulated mechanisms.

Copyright © 2009 S. Karger AG, Basel

Although several candidate genes have been studied in association with human performance, none has created as much debate, or consumed as many publication pages, as an angiotensin-converting enzyme (ACE) insertion-deletion polymorphism in intron 16. The particular polymorphism is characterised by the presence (insertion, I-allele) or absence (deletion, D-allele) of a 287-bp *Alu* repeat sequence [1]. We all carry two versions of the ACE-allele and in 1,906 British males free from cardiovascular disease the presence of II, ID and DD genotypes has been reported as 0.24, 0.50 and 0.26, respectively (I-allele frequency = 0.49) [2]. In Caucasian Europeans, the ACE genotype frequencies for II, ID and DD genotypes are generally found in the ratio of 1:2:1 [3]. However, the ratio of genotypes carried does vary depending on the population studied. There is a trend toward a higher frequency of the D-allele in Nigerians [3], whereas in Samoans and Yanomami Indians there is a much higher frequency of the I-allele [3], and in Pima Indians the I-allele frequency is as high as

0.71 [4]. Within the UK, there are also significant differences with a higher frequency of the D-allele in Afro-Caribbean people [5] and a much higher frequency of the II genotype in UK-domiciled South Asians than Caucasians [6].

Although the ACE I/D polymorphism occurs in an intron, it is an exceptionally strong and consistent marker for ACE activity in many different Caucasian populations [7–13]. ACE is consistently highest in the DD subjects, intermediate in the ID and lowest in the II subjects. Although the polymorphism accounted for 18–20% of the variation in ACE in Caucasians of the Ectim study [7], 30–40% in Scandinavian Caucasians [10] and 47% in the Caucasians studied by Rigat [1], there are marked racial differences. In Pima Indians, the ACE I/D polymorphism accounts for only 6.5% of the variation in serum ACE activity [4], 13% in Kenyans [14] and there is no relationship apparent at all in African-Americans [15] or black South Africans [16]. The variance in the proportion of ACE activity associated with the ACE I/D polymorphism in different populations may suggest that there are racial differences in the transcriptional regulation of ACE and that the association with ACE activity is mediated through linkage disequilibrium with an as yet unknown neighbouring functional quantitative trait locus. Nonetheless, in European Caucasian populations it remains a strong and consistent marker for variation in ACE activity. There are also intrinsic differences in the renin-angiotensin system (RAS) between racial groups which make direct comparison between studies and extrapolation from studies of Caucasian subjects difficult. A lower plasma rennin activity is typically found in Black people compared with Caucasians [17, 18], and a lower plasma aldosterone level is found in Black children [19].

It is this racial difference in the RAS coupled with differences in gene frequency and the association between the ACE I/D polymorphism and ACE levels that may be one of the important keys to interpreting the different results from various studies.

The ACE I/D Polymorphism and Human Performance at Sea Level: Cross-Sectional Studies

Just over 10 years ago, the I-allele was associated with endurance performance, initially being found with excess frequency in elite distance runners [20] and rowers [21]. All of the 64 Caucasian (43 male, 21 female) Australian national rowers attending their pre-Olympics selection trial were examined for their frequency of the ACE I/D polymorphism [21]. They were a nationally representative group of elite rowers with selection bias reduced in view of their national identification programme screening for talented individuals (41 of the cohort won more rowing medals in the 1996 Olympics than any other nation). In comparison with a well-matched control group there was a clear excess of the I-allele and the II genotype reported.

The following year, a study that had targeted potential Olympic athletes was published [20]. 495 potential Olympic competitors formed the cohort. Of this total

91 were runners over distances from 100 m to 100 km, and in comparison with a large control group there was a significant excess of both the I-allele and II genotype reported due to a skew toward the I-allele in the ≥5,000 m group. Examination of the relative distribution of alleles by duration of event (≥200 m mainly anaerobic, 400–3,000 m mixed aerobic and anaerobic, and ≥5,000 m predominantly aerobic or endurance) was informative. There was a significant linear trend of increasing I-allele frequency with distance run with the proportion of I-alleles increasing from 0.35 to 0.53 and 0.62 amongst those running ≤200, 400–3,000 and ≥5,000 m, respectively. There was also a skew in favour of the D-allele amongst the sprinters (D-allele frequency 0.62 in those running ≤200 m). This was the first study to suggest that the ACE I/D polymorphism was not simply a ubiquitous 'gene marker of performance' but that the I-allele may confer a benefit in more endurance-related events and the D-allele in more power-oriented events.

At around the same time, other workers found, and published, no such association between the ACE I/D polymorphism and elite human performance [22–24]. Taylor et al. [23] examined 120 Australian national athletes from sports deemed to demand a high level of aerobic fitness (26 hockey players, 25 cyclists, 21 skiers, 15 track and field athletes, 13 swimmers, 7 rowers, 5 gymnasts and 8 'other'). They found no difference in ACE genotype and allele frequency compared with controls. Similarly, the cohort examined by Karjalainen et al. [22] of 80 elite endurance athletes from Finnish national teams included the various disciplines of long distance running, orienteering, cross-country skiing and the triathlon. Again, they found no difference in the ACE I/D polymorphism between the athletes and controls. One of the largest (n = 192) studies to date of national and international athletes (cross-country skiing, biathlon, Nordic combined, long distance running, middle distance running and road cycling) was also negative [24].

However, a key consideration of genetic association studies is case definition. The common denominator among these early negative studies was the inclusion of athletes from mixed sporting disciplines, thereby producing phenotypic heterogeneity and obscuring any possible genetic association.

To clarify this point, a study specifically addressed this issue [25]. A mixed cohort of 217 Russian athletes of regional or national standard were recruited from the following sports: swimming (n = 66), track and field athletics (n = 81), cross-country skiing (n = 52), and the triathlon (n = 18). The a priori intention was to clarify whether previously documented associations between the ACE polymorphism and elite athletes found in the study of a single sporting discipline could be duplicated in Russian athletes. Secondly, to determine whether any association would be lost if athletes from sports with different elements of power and endurance activity were combined. The athletes were stratified into groups according to event duration, covering a spectrum from the more 'power'-oriented to the more endurance-oriented. Short distance athletes (SDAs) consisted of 'sprinters' performing their given task in under 1 min; middle distance athletes (MDAs) were those competing over 1–20 min, and long distance athletes were those performing over more than 20 min. The athletes were further classified by their standard of competition based on previous performances. 'Outstanding' athletes (n = 141) were at least national representatives and included 81 European and Russian champions and 19 Olympic or World champions. 'Average' athletes (n = 76) were regional competitors. Selection bias was reduced by employing multiple methods of recruitment.

Compared with a control group of 449 Caucasian Russians, the genotype distribution and allele frequency amongst the whole cohort (19.3% II, 51.2% ID, 29.5% DD, D-allele frequency 0.55) and that of outstanding athletes alone was similar to that amongst sedentary controls. In the cohort of outstanding athletes, ACE I/D polymorphism was indeed associated with event duration: an excess of D-alleles being noted in the SDA group, and an elevated frequency of the I-allele in the MDA group. There was no increasing linear trend of the I-allele with increasing aerobic component as has previously been described [20]. This may relate to differences in categorizing athletes. Myerson examined allele frequency by distance ran, whereas the study of Nazarov et al. [25] examined allele frequency by duration of event in order to compare outstanding athletes of different disciplines. As such, a 5,000 m runner or skier would be categorized as an MDA by the time criteria, but a long distance competitor by the distance criteria of Myerson and colleagues. Nevertheless, both groups (≥5,000 m and MDAs) show an excess of the I-allele, with similar frequencies (0.63 and 0.62, respectively).

These data support the hypothesis that the study of a mixed cohort of athletes from various sporting disciplines may obscure any association between elite athletes and the ACE I/D polymorphism. It is also important to note that if non-elite athletes are studied, association is often lost. The track and field athletes in the study of Nazarov et al. [25] demonstrated an excess of the D-allele in outstanding SDAs which, was not present in the average athletes. Similar findings have been demonstrated in other cohorts of athletes from different events. In a study of 126 Italian 'candidate Olympic athletes' when anaerobic-based and aerobic-based events were included together, there was no difference in genotype or allele distribution compared with a control group. However, although small numbers, comparing aerobic-based events (n = 33) with anaerobic-based events (n = 19) there was an excess of the II genotype in the former [26]. A further report in 139 Korean male elite athletes from various sporting disciplines (including basketball, soccer, baseball, gymnastics, volleyball, long-distance running, judo and ice hockey) also failed to show a difference in ACE I/D polymorphism frequencies compared with control (although there was a trend to excess of the I-allele in long distance runners) [27]. Indeed, if one looks back at the original study by Myerson [20] in UK Olympic athletes, it was also demonstrated that in combining 404 athletes from diverse disciplines such as judo and pole vaulting, there was no difference in allele frequency compared to the control population [20]. There has been however, the occasional report of an excess of the I-allele in mixed cohorts of athletes [28].

Subsequent studies examining cohorts of elite athletes from single sporting disciplines have also found an excess of the II genotype, for example in endurance runners versus soccer players [29] and Olympic class Spanish runners versus a control group [30].

That said, however, it has to be acknowledged that even in studies examining a single sporting discipline, associations have not always been found. The ACE I/D polymorphism was not found to relate to endurance athlete status in 221 national and 70 international Kenyan athletes [14]. However, the ACE I/D polymorphism only explains 13% of the variance in ACE activity in a general Kenyan population [14]. If the putative mechanism of effect is via differences in ACE activity and downstream effects within the RAS, then the degree of association between genotype and ACE activity could be crucial. Similarly, other studies have even found diametrically opposed results, for example a relative lack of the II genotype in rowers [31].

Although not strictly a single sporting discipline, few would argue that endurance is a critical factor in Ironman triathlons. In participants of the South African Ironman Triathlon a significantly higher frequency of the I-allele in the fastest 100 South African-born finishers (103 I, 51.5%, and 97 D, 48.5%) compared with the 166 South African-born control subjects (140 I, 42.2%, and 192 D, 57.8%) has been reported [32]. There was also a significant linear trend for the allele distribution among the fastest 100 finishers (I-allele = 51.5%), slowest 100 finishers (I-allele = 47.5%), and control subjects (I-allele = 42.2%; p = 0.033). Similarly, a study of 215 marathon runners found an increased frequency of the II genotype in the fastest runners [33].

It is important to realize that the ACE I/D polymorphism should not be considered a 'gene for human performance', but a marker that one would expect association only in the truly elite athlete according to the nature of the specific event. There has been an excess of the D-allele noted amongst sprinters (*D*-allele frequency 0.62 in those running ≤200 m) [20] and elite athletes completing their event in <1 min [25]. This relationship between the D-allele and power sports occurring over a short time period is distinct from that of the I-allele and enhanced endurance performance. If a discipline is spread over varying distances (such as swimming, running) then cohort stratification should be considered according to anaerobic/aerobic or power/endurance component.

Swimmers are a particularly interesting example. A strong relationship exists between upper body anaerobic power and performance in swim events from 50 to 400 m [34, 35] and upper body arm power is a highly significant performance factor in the longer swimming distances [35].

Accordingly, one study in European and Commonwealth championship Caucasian swimmers subdivided them into successful competitors over <400 and >400 m. A significant excess of the D-allele and DD genotype compared with large control groups was reported due to an excess of the D-allele and DD genotype in those swimming over the shorter distances [36]. A small study (n = 35) of elite very long distance swimmers classified as better at 1- to 10-km distances (n = 19) or those best at 25-km races (n = 16) again revealed an apparent excess of the D-allele in those better at

Table 1. Chronological summary of population association studies examining the relationship between ACE I/D polymorphism and human performance by study first author

Study first author	Subjects	Analysis by single sporting discipline?	Association with the ACE I/D polymorphism
Montgomery [1998]	25	yes, elite climbers	yes
Myerson [1999]	91	yes, runners	yes
Gayagay [1998]	64	yes, rowers	yes
Taylor [1999]	120	no, multiple disciplines	no
Karjalainen [1999]	80	no, multiple disciplines	no
Rankinen [2000]	192	no, multiple disciplines	no
Alvarez [2000]	60	no, multiple disciplines	yes
Woods [2001]	56	yes, swimmers	yes
Nazarov [2001]	141	no, but stratified by event duration	yes
Scanavini [2002]	126	no, multiple disciplines	no in whole cohort but yes in runners
Tsianos [2004]	35	yes, swimmers	yes
Oh [2007]	139	no, multiple disciplines	no
Lucía [2005]	27	yes, runners	yes
Collins [2004]	100	no (triathlon)	yes
Scott [2005]	291	yes, runners	no, but weak relationship between genotype and ACE activity in Kenyans
Hruskovicová [2006]	215	yes, runners	yes
Muniesa [2008]	141	no, multiple disciplines	yes, reduced I-allele in rowers
Juffer [2009]	52	yes, runners	yes

shorter distance swimming and an excess of the I-allele in those swimming ultra-endurance distances [37]. While a 1- to 10-km swim could never be considered a short event, perhaps the upper body arm power referred to by Hawley et al. [35] has primacy unless the event is a really long endurance swim. Sadly, the numbers are too small to reveal any resounding insight.

Myerson et al. [20] had also found a significant excess of the D-allele among elite swimmers. Interestingly, Nazarov et al. [25] found again that a mixed cohort of average and elite swimmers demonstrated no skew in genotype distribution but examining allele frequencies by duration of event in elite athletes only there was a significant excess of the D-allele in swimmers completing their event <1 min. Diluting a cohort with non-elite athletes may result in failure to find an association with the ACE I/D polymorphism. To test whether a genetic marker occurs more frequently in elite athletes than controls requires a homogenous cohort from the same sporting discipline with large, matched control groups (table 1).

The ACE I/D Polymorphism and Prospective Training Studies at Sea Level

To examine the effect of a gene polymorphism that may have only small, but significant, effects on human performance and to attempt to tease this effect out from the multifactorial make-up of an elite athlete often requires examining a period of gene-environment interaction. Most often, this involves studying the influence of a period of training on a performance phenotype.

The I-Allele and Endurance

The earliest prospective study examining the influence of the ACE I/D polymorphism on performance parameters was published 11 years ago. While perhaps not employing a gold standard method of assessing endurance, it did reveal a genotype-dependent difference in a performance parameter with training. Duration of loaded repetitive biceps flexion was independent of genotype at baseline in 78 British Army recruits but was found to increase 11-fold more amongst those of the II genotype compared to the DD genotype after a 10-week general physical training programme [38]. Similarly, in female non-elite Caucasian Turkish athletes over a 6-week training regimen, the II genotype related to better improvements in medium duration aerobic endurance performance (such that a significant linear trend: II > ID > DD occurred) [39].

A study in American recruits does not support these findings [40]. However, these authors specifically aimed to assess whether an association between the ACE I/D polymorphism and performance phenotypes with training existed in a heterogeneous, mixed-sex, multi-ethnic population of US military recruits. 147 recruits were studied before and after 8 weeks of basic training which, in marked contrast to UK recruits, is partly performed in streamed quartiles according to baseline fitness. In addition, the performance marker used was the Army Physical Fitness Test which the US Army itself considers to be a measure of combined endurance and strength. The application of a heterogeneous training programme to a heterogeneous cohort as might be expected did not reveal any significant ACE I/D polymorphism association. In addition, if the effect of the polymorphism is mediated by differences in ACE activity, it is important to remember that there is no such relationship in African-Americans, a significant portion of this cohort. It is more difficult to reconcile data from a study in Turkish Army recruits that found a performance enhancement after 6 months training related to the D-allele over a 2,400-metre running distance [41].

The D-Allele and Power-Oriented Performance

One report documents the effects of 9 weeks of strength training on quadriceps muscle strength in 33 healthy male volunteers [42]. A significant interaction between ACE genotype and isometric training occurred with greater strength gains shown by subjects with the D-allele. This increase in strength may contribute to the apparent excess of the D-allele among power-oriented events.

A further study subsequently has also demonstrated an apparent improvement in strength response to 6 weeks of training associated with the DD genotype in 99 Caucasian male non-elite athletes [43]. Similarly, in female non-elite Caucasian Turkish athletes over a 6 week training regimen, the DD genotype conferred a more advantageous performance enhancement in shorter duration and higher intensity endurance activities [39]. In support of these studies, although in 213 older obese people, an 18-month training programme of walking and light weight lifting resulted in individuals with the DD genotype showing greater gains in knee extensor strength compared with II individuals [44].

Although involving a shorter period of training, another cohort of elderly subjects showed that the DD genotype was associated with greater quadriceps muscle volume at baseline but did not influence their knee extensor muscle strength [45]. Peak elbow flexor strength in a large cohort of 631 subjects after training 2 days a week for 12 weeks was greater in the contralateral but not the trained arm in association with the DD genotype [46]. Other studies have found no improvement in elbow flexor strength associated with the ACE I/D polymorphism after training [47].

The ACE I/D Polymorphism and Human Performance at High Altitude

It was also noted over 10 years ago that in the extreme physiological environment of high altitude (HA) mountaineers demonstrate an allele skew compared to a normal population [20]. In 25 elite male British mountaineers who had ascended beyond 7,000 m without the use of supplemental oxygen there was a significant excess of the II genotype and I-allele with no DD genotypes in those climbers who had gone beyond 8,000 m without oxygen. This is an incredibly small group for a gene association study but we are limited to an extent by the relatively small numbers enduring such physiological stress. Again, despite the numbers, this report spawned a plethora of subsequent studies seeking to confirm or refute an association, and to examine potential mechanisms.

Subsequent work also reported a relative increased frequency of the I-allele (0.55 vs. 0.36, successful vs. unsuccessful, respectively) in those climbers who had reached 8,000 m, from a cohort of 139 who had attempted to do so [48]. Similarly, from a cohort of 284, success at climbing Mont Blanc (4,807 m) has been associated with the I-allele [49].

The ACE I/D Polymorphism and Human Performance: Potential Mechanisms Involved

The ACE polymorphism not only influences circulating ACE activity, it also appears to be a determinant of ACE at a cellular level [50–53]. This may have downstream effects in the production of angiotensin II and breakdown of bradykinin that will influence various factors including blood flow, substrate utilization and cell growth relevant to human performance.

In support of this, DD subjects demonstrate greater plasma angiotensin II concentrations following angiotensin I infusion [54, 55], and the contractile response of internal mammary arteries to angiotensin I and II suggest that angiotensin I conversion is greatest in the presence of the D-allele [56]. Local tissue conversion of angiotensin I may also be relevant since DD subjects have a significantly enhanced forearm vasoconstrictor response to angiotensin I infusion that is not accompanied by differences in serum angiotensin II levels [57].

It is important to note that angiotensin II type 1 receptor (AT_1)-mediated angiotensin II stimulation is crucial for optimal overload-induced skeletal muscle hypertrophy [58]. Further, it has been demonstrated that AT_1 blockade in rats undergoing eccentric contraction training over 4 weeks prevents the training-induced increase in muscle mass and muscle contractile force compared with controls [59]. Angiotensin II is also known to increase the contraction-induced oxygen uptake and the tension during tetanic stimulation in rats [60]. This effect may come at a cost, and it is known that angiotensin II administration in rats reduces metabolic efficiency with skeletal muscle wasting [61]. The redirection of blood flow from type I muscle fibres to the type II fibres that are favoured in power performance (a sprinter may have 80% fast twitch fibres, an endurance athlete 20%) is also influenced by angiotensin II [60] and would be potentially beneficial in power-oriented events.

Other potential mechanisms by which varied angiotensin II levels could influence human performance are via its effect as a direct stimulator of cellular growth (both hypertrophic and hyperplastic); the induction of various endogenous growth factors and the facilitation of sympathetic nerve transmission by enhancing noradrenaline release from peripheral sympathetic nerve terminals and the CNS [reviewed in 62].

There is also evidence of reduced bradykinin degradation in II subjects with low ACE activity [54]. Reduced bradykinin degradation may favourably alter substrate metabolism in II subjects with improvements in the efficiency of skeletal muscle. Indeed, a bradykinin-mediated increase in nitric oxide production in skeletal muscle facilitates glucose uptake [63, 64]. Muscle work increases muscle blood flow (and bradykinin in human venous effluent) [65, 66] with increased glucose uptake in humans, an effect reproduced by bradykinin infusion in the human forearm that can be blocked by inhibiting bradykinin-generating proteases [67]. Further, flow-mediated vasodilatation has been reported to be ACE I/D polymorphism dependent with greater flow-mediated vasodilatation in II subjects compared to DD subjects [68]. This may represent the effect of an increased half-life of the vasodilator bradykinin associated with the I-allele and lower ACE activity as opposed to the blunted nitric oxide vasodilatory response seen in the forearm vessels of DD subjects [69].

It can be seen therefore that the effect of bradykinin on muscle blood flow and substrate utilization is a key factor. Indeed, the oxidative capacity of muscle (greatest in the type 1 fibres that correlate with efficiency in cyclists) [70] is in direct correlation and may be interdependent with muscle kallikrein from which bradykinin originates [71].

In addition, a reduction in ACE activity leads not only to an attenuation of angiotensin II production and a decrease in bradykinin degradation but also to an increase in Ang-(1–7) [72]. This may further favour vasodilation and potential substrate delivery as the actions of Ang-(1–7) are most often opposite those of angiotensin II.

The important consideration however is whether differences in ACE, angiotensin II, bradykinin and Ang-(1–7) translate measurable factors that may be relevant in human performance.

A high level of aerobic fitness is an important requirement, but not a final determinant, of elite endurance performance. The standard measure of aerobic capacity, VO_{2max}, is influenced by genetic and environmental factors with marked inter-individual response to training. As such, in the search for a physiological link between the ACE I/D polymorphism and elite human performance, the study of central cardiorespiratory factors such as VO_{2max} was a natural avenue of investigation. As the most important environmental factor in determining VO_{2max} is regular physical activity, it is vital that any genetic study is charged with standardizing the training protocol within, and between, groups; however, despite several studies concluding that there has been no consistent effect of the ACE I/D polymorphism on VO_{2max} reported [40, 73, 74].

Hagberg et al. [75] did however find an increase in VO_{2max} associated with the I-allele that was entirely due to an increase in maximal arteriovenous oxygen difference (rather than increased in cardiac output). The effect of angiotensin II and bradykinin on local skeletal muscle bioenergetics may provide the key to unlocking a potential peripheral muscle effect rather than a central cardiopulmonary effect.

One report supports that notion. 58 British military recruits had their VO_{2max} and delta efficiency (delta work accomplished min^{-1} divided by delta energy expended min^{-1}, expressed as a percentage) assessed before and after 11 weeks of physical training [76]. VO_{2max} was no different before or after training according to ACE I/D polymorphism, but delta efficiency (independent of genotype at baseline) rose significantly with training only amongst those of II genotype (absolute change 1.87% for II subjects, –0.26% for DD), a proportional increase in efficiency of 8.62%. This would certainly be enough to have a significant biological impact. This study is supported by the greater peripheral tissue oxygenation and lesser rise in lactate, reflecting greater muscle efficiency, that occurs in II compared with DD subjects during exercise in patients with chronic airway disease [77] and the greater aerobic skeletal muscle efficiency found in Chinese chest patients with the II genotype [78], although not in 62 untrained sedentary females [79]. The study in military recruits has the advantage in maintaining other environmental factors as constants (subjects were sleeping in the same location for a similar duration, eating the same diet and exercising to identical supervised targets at identical times as well as being of one sex and from a very narrow age band), which helps to elucidate any subtle gene effect.

Subsequent work has offered further insight into this improvement in muscle efficiency. In 41 healthy young volunteers, skeletal muscle biopsies from the vastus lateralis have demonstrated that II subjects have a higher percentage of slow-twitch type I fibres (more efficient than fast-twitch fibres at low velocities of contraction). In addition, II subjects had a lower percentage of fast-twitch type IIb fibres than DD subjects with a linear trend for a decrease in type I fibres and increase in type IIb fibres from the II through the ID to the DD genotypes [80]. This may be a mechanism that explains the physiological link between the I-allele and improved endurance performance by means of improving muscle efficiency, and correlates with the fact that the oxidative capacity of muscle is greatest in type 1 fibres (that are more common in endurance runners and correlate with efficiency in cyclists) [70].

Possible Mechanisms Involved at High Altitude
Numerous studies have explored possible explanations as to why the I-allele may confer a performance advantage at HA. Obviously, a reduction in any acute mountain sickness (AMS) would be beneficial at HA. However, three studies have shown no apparent benefit related to the ACE genotype in terms of AMS, one cohort of 284 subjects [49], another of 151 [81] and another of 103 Nepalese [82]. Similarly, there is no genotype-dependent difference in HA pulmonary oedema [81, 83]. Other mechanisms have therefore been explored. One study examined an obvious factor, oxygen saturation, at HA [84]. Although again relatively small in subject numbers due to the nature of the activity, two groups of Caucasian subjects were studied. SaO_2 with ascent was significantly associated with the ACE genotype (SaO_2 at 3,870 m was 85.4, 87.5, 90.1% for DD, ID and II genotypes, respectively) in a rapid ascent group (n = 32) on Mount Everest with a relatively sustained SaO_2 in the II subjects. Whereas there was no similar association found with a slow ascent group on Kanchenjunga. This could possibly be explained by time-dependent altitude-related changes in the RAS.

This earlier work is supported by a more recent study in a Peruvian cohort [85]. In 142 subjects of mainly Quechua origin, resting and exercise SaO_2 was strongly associated with the ACE genotype at HA with approximately 4% of the total variance in SaO_2 being attributable to the ACE I/D polymorphism. II subjects maintained an SaO_2 around 2.3% higher than ID or DD subjects, regardless of whether the subjects were born at high or low altitude.

At altitude, diminished alveolar oxygen tensions lead to lower arterial oxygen tensions. However, the fall in oxygen availability with altitude is sensed by hypoxia-sensitive AT_1 receptors in the carotid bodies [86] which stimulate an increase in minute ventilation. This hypoxic ventilatory response helps sustain arterial oxygen saturations and tissue oxygen delivery and demonstrates a marked inter-individual variation. Patel et al. [87] have demonstrated from a cohort of 60 subjects that the hypoxia-induced percentage rise in minute ventilation is strongly and significantly associated with the II genotype (II 39.6 ± 4.1% vs. DD 28.4 ± 2.2%).

Conclusions

With the caveat of the frequent small cohorts that we must accept by the very nature of the elite groups being studied, there nevertheless remains a significant breadth and depth of evidence supporting a role for the ACE I/D polymorphism in human performance. There will never be a cohort of thirty thousand elite athletes from the same discipline. There will always be controversies with the interpretation of association studies. Attempting to reduce selection bias and giving due consideration to appropriate case definition as well as studying athletes from homogenous sporting disciplines with appropriately large control groups offer the best chance of reliable, reproducible outcomes. In concert with increasing data regarding an influence of the polymorphism on plausible mechanistic pathways, the balance of the evidence currently is in favour of there being a small but significant influence of the ACE I/D polymorphism on human performance. That small difference, following a suitable period of training and therefore gene-environment interaction, could be a significant factor between success and failure.

The ACE I/D polymorphism should not be considered a 'gene for human performance', but a marker of modulation such that one would expect an excess of the I-allele in the truly elite endurance athlete, with a concordant excess of the D-allele represented in the more power-oriented events. Of course, because elite athletes are still 'made' not 'born' there will always be elite athletes in endurance events with the DD genotype.

In what direction should the next generation of studies proceed? With particular reference to the association of the I-allele with elite endurance performance then a move from genetic to pharmacological studies could prove very revealing. If the low ACE activity associated with the I-allele has key physiological importance in the response to training, then studies using ACE inhibitor drugs to reduce the effect of ACE or angiotensin receptor blocker drugs (that only interfere with the action of angiotensin II and do not prolong the half-life of bradykinin) could provide great insight. Indeed, some commentators are already concerned that these drugs may, or have become, drugs of abuse in sport [88].

References

1 Rigat B, Hubert C, Alhencgelas F, Cambien F, Corvol P, Soubrier F: An insertion deletion polymorphism in the angiotensin I-converting enzyme gene accounting for half the variance of serum enzyme levels. J Clin Invest 1990;86:1343–1346.

2 Miller GJ, Bauer KA, Barzegar S, Cooper JA, Rosenberg RD: Increased activation of the haemostatic system in men at high risk of fatal coronary heart disease. Thrombosis Haemost 1996;75:767–771.

3 Barley J, Blackwood A, Carter ND, Crews DE, Cruickdshank JK, Jeffery S, Ogunlesi AO, Sagnella GA: Angiotensin converting enzyme insertion/deletion polymorphism: association with ethnic origin. J Hypertens 1994;12:955–957.

4 Foy C, McCormack LJ, Knowler WC, Barrett JH, Catto A, Grant PJ: The angiotensin-1 converting enzyme (ACE) gene I/D polymorphism and ACE levels in Pima Indians. J Med Genet 1996;33:336–337.

5 Barley J, Blackwood A, Miller M, Markandu ND, Carter ND, Jeffery S, Cappuccio FP, MacGregor GA, Sagnella GA: Angiotensin converting enzyme gene I/D polymorphism, blood pressure and the renin-angiotensin system in Caucasian and Afro-Caribbean peoples. J Hum Hypertens 1996;10:31–35.

6 Sagnella GA, Rothwell MJ, Onipinla AK, et al: A population study of ethnic variations in the angiotensin-converting enzyme I/D polymorphism: relationships with gender, hypertension and impaired glucose metabolism. J Hypertens 1999;17:657–664.

7 Cambien F, Costerousse O, Tiret L, Poirier O, Lecerf L, Gonzales MF, Evans A, Arveiler D, Cambou JP, Luc G, Rakotovao R, Ducimetiere P, Soubrier F, Alhenc-Gelas F: Plasma level and gene polymorphism of angiotensin-converting enzyme in relation to myocardial infarction. Circulation 1994;90:669–676.

8 Busjahn A, Knoblauch H, Knoblauch M, Bohlender J, Menz M, Faulhaber HD, Becker A, Schuster H, Luft FC: Angiotensin-converting enzyme and angiotensinogen gene polymorphisms, plasma levels, cardiac dimensions: a twin study. Hypertension 1997;29:165–170.

9 Danser AH, Derkx FH, Hense HW, Jeunemaitre X, Riegger GA, Schunkert H: Angiotensinogen (M235T) and angiotensin-converting enzyme (I/D) polymorphisms in association with plasma renin and prorenin levels. J Hypertens 1998;16:1879–1883.

10 Agerholm-Larsen B, Tybjserg-Hansen A, Schnohn P, Nordestgaard BG: ACE gene polymorphism explains 30–40% of variability in serum ACE activity in both women and men in the population at large: the Copenhagen City Heart study. Atherosclerosis 1999;147:425–427.

11 Kohno M, Yokokawa K, Minami, M, Kano H, Yasunari K, Hanehira T, Yoshikawa J: Association between angiotensin-converting enzyme gene polymorphisms and regression of left ventricular hypertrophy in patients treated with angiotensin-converting enzyme inhibitors. Am J Med 1999;106:544–549.

12 Rossi GP, Narkiewicz K, Cesari M, Winnicki M, Bigda J, Chrostowska M, Szczech R, Pawlowski R, Pessina AC: Genetic determinants of plasma ACE and renin activity in young normotensive twins. J Hypertens 1999;17:647–655.

13 Martinez EPA, Escribano J, Sanchis C, Carrion L, Artigao M, Divison JA, Masso J, Vidal A, Fernandez JA: Angiotensin-converting enzyme (ACE) gene polymorphisms, serum ACE activity and blood pressure in a Spanish-Mediterranean population. J Hum Hypertens 2000;14:131–135.

14 Scott RA, Moran C, Wilson RH, Onywera V, Boit MK, Goodwin WH, Gohlke P, Payne J, Montgomery H, Pitsiladis YP: No association between Angiotensin Converting Enzyme (ACE) gene variation and endurance athlete status in Kenyans. Comp Biochem Physiol A Mol Integr Physiol 2005;141:169–175.

15 Bloem LJ, Manatunga AK, Pratt JH: Racial difference in the relationship of an angiotensin I- converting enzyme gene polymorphism to serum angiotensin I- converting enzyme activity. Hypertension 1996;27:62–66.

16 Payne JR, Dhamrait SS, Gohlke P, Cooper J, Scott RA, Pitsiladis YP, Humphries SE, Rayner B, Montgomery HE: The impact of ACE genotype on serum ACE activity in a black South African male population. Ann Hum Genet 2007;71:1–7.

17 Brunner HR, Laragh JH, Baer L: Essential hypertension: renin and aldosterone, heart attack and stroke. N Engl J Med 1972;286:441–449.

18 Kotchen TA, Guthrie GP Jr, Cottrill CM, McKean HE, Kotchen JM: Low renin-aldosterone in 'prehypertensive' young adults. J Clin Endocrinol Metab 1982;54:808–814.

19 Pratt JH, Jones JJ, Miller JZ, Wagner MA, Fineberg NS: Racial differences in aldosterone excretion and plasma aldosterone concentrations in children. N Engl J Med 1989;321:1152–1157.

20 Myerson S, Hemingway H, Budget R, et al: Human angiotensin I-converting enzyme gene and endurance performance. J Appl Physiol 1999;87:1313–1316.

21 Gayagay G, Yu B, Hambly B, Boston T, Hahn A, Celermajer DS, Trent RJ: Elite endurance athletes and the ACE I allele – the role of genes in athletic performance. Hum Genet 1998;103:48–50.

22 Karjalainen J, Kujala UM, Stolt A, Mantysaari M, Viitasalo M, Kainulainen K, Kontula K: Angiotensinogen gene M235T polymorphism predicts left ventricular hypertrophy in endurance athletes. J Am Coll Cardiol 1999;34:494–499.

23 Taylor RR, Mamotte CDS, Fallon K, Bockxmeer FM: Elite athletes and the gene for angiotensin-converting enzyme. J Appl Physiol 1999;87:1035–1037.

24 Rankinen T, Wolfarth B, Simoneau J, et al: No association between the angiotensin-converting enzyme ID polymorphism and elite endurance athlete status. J Appl Physiol 2000;88:1571–1575.

25 Nazarov IB, Woods DR, Montgomery HE, et al: The angiotensin converting enzyme I/D polymorphism in Russian athletes. Eur J Hum Genet 2001;9:797–801.

26 Scanavini D, Bernardi F, Castoldi E, Conconi F, Mazzoni G: Increased frequency of the homozygous II ACE genotype in Italian Olympic endurance athletes. Eur J Hum Gen 2002;10:576–577.

27 Oh SD: The distribution of I/D polymorphism in the ACE gene among Korean male elite athletes. J Sports Med Phys Fitness 2007;47:250–254.

28 Alvarez R, Terrados N, Ortolano R, Iglesias-Cubero G, Reguero JR, Batalla A, Cortina A, Fernandez-Garcia B, Rodriguez C, Braga S, Alvarez V, Coto E: Genetic variation in the renin-angiotensin system and athletic performance. Eur J Appl Physiol 2000; 82:117–120.

29 Juffer P, Furrer R, González-Freire M, Santiago C, Verde Z, Serratosa L, Morate FJ, Rubio JC, Martin MA, Ruiz JR, Arenas J, Gómez-Gallego F, Lucia A: Genotype distributions in top-level soccer players: a role for ACE? Int J Sports Med 2009, Epub ahead of print.

30 Lucía A, Gómez-Gallego F, Chicharro JL, Hoyos J, Celaya K, Córdova A, Villa G, Alonso JM, Barriopedro M, Pérez M, Earnest CP: Is there an association between ACE and CKMM polymorphisms and cycling performance status during 3-week races? Int J Sports Med 2005;26:442–447.

31 Muniesa CA, González-Freire M, Santiago C, Lao JI, Buxens A, Rubio JC, Martín MA, Arenas J, Gomez-Gallego F, Lucia A: World-class performance in lightweight rowing: is it genetically influenced? A comparison with cyclists, runners and non-athletes. Br J Sports Med 2008, Epub ahead of print.

32 Collins M, Xenophontos SL, Cariolou MA, Mokone GG, Hudson DE, Anastasiades L, Noakes TD: The ACE gene and endurance performance during the South African Ironman Triathlons. Med Sci Sports Exerc 2004;36:1314–1320.

33 Hruskovicová H, Dzurenková D, Selingerová M, Bohus B, Timkanicová B, Kovács L: The angiotensin converting enzyme I/D polymorphism in long distance runners. J Sports Med Phys Fitness 2006;46: 509–513.

34 Hawley JA, Williams MM: Relationship between upper body anaerobic power and freestyle swimming performance. Int J Sports Med 1991;12:1–5.

35 Hawley JA, Williams MM, Vickovic MM, Handcock PJ: Muscle power predicts freestyle swimming performance. Br J Sports Med 1992;26:151–155.

36 Woods DR, Hickmann M, Jamshidi Y, et al: Elite swimmers and the D allele of the ACE I/D polymorphism. Hum Genet 2001;108:230–232.

37 Tsianos G, Sanders J, Dhamrait S, Humphries S, Grant S, Montgomery H: The ACE gene insertion/deletion polymorphism and elite endurance swimming. Eur J Appl Physiol 2004;92:360–362.

38 Montgomery HE, Marshall RM, Hemingway H, et al: Human gene for physical performance. Nature 1998;393:221–222.

39 Cam S, Colakoglu M, Colakoglu S, Sekuri C, Berdeli A: ACE I/D gene polymorphism and aerobic endurance development in response to training in a non-elite female cohort. J Sports Med Phys Fitness 2007; 47:234–238.

40 Sonna LA, Sharp MA, Knapik JJ, et al: Angiotensin-converting enzyme genotype and physical performance during US Army basic training. J Appl Physiol 2001;91:1355–1363.

41 Cerit M, Colakoglu M, Erdogan M, Berdeli A, Cam FS: Relationship between ACE genotype and short duration aerobic performance development. Eur J Appl Physiol 2006;98:461–465.

42 Folland J, Leach B, Little T, Hawker K, Myerson S, Montgomery H, Jones D: Angiotensin-converting enzyme genotype affects the response of human skeletal muscle to functional overload. Exp Physiol 2000;85:575–579.

43 Colakoglu M, Cam FS, Kayitken B, Cetinoz F, Colakoglu S, Turkmen M, Sayin M: ACE genotype may have an effect on single versus multiple set preferences in strength training. Eur J Appl Physiol 2005;95:20–26.

44 Giaccaglia V, Nicklas B, Kritchevsky S, Mychalecky J, Messier S, Bleecker E, Pahor M: Interaction between angiotensin converting enzyme insertion/deletion genotype and exercise training on knee extensor strength in older individuals. Int J Sports Med 2008;29:40–44.

45 Charbonneau DE, Hanson ED, Ludlow AT, Delmonico MJ, Hurley BF, Roth SM: ACE genotype and the muscle hypertrophic and strength responses to strength training. Med Sci Sports Exerc 2008; 40:677–683.

46 Pescatello LS, Kostek MA, Gordish-Dressman H, Thompson PD, Seip RL, Price TB, Angelopoulos TJ, Clarkson PM, Gordon PM, Moyna NM, Visich PS, Zoeller RF, Devaney JM, Hoffman EP: ACE ID genotype and the muscle strength and size response to unilateral resistance training. Med Sci Sports Exerc 2006;38:1074–1081.

47 Thomis MA, Huygens W, Heuninckx S, Chagnon M, Maes HH, Claessens AL, Vlietinck R, Bouchard C, Beunen GP: Exploration of myostatin polymorphisms and the angiotensin-converting enzyme insertion/deletion genotype in responses of human muscle to strength training. Eur J Appl Physiol 2004;92:267–274.

48 Thompson J, Raitt J, Hutchings L, Drenos F, Bjargo E, Loset A, Grocott M, Montgomery H, Caudwell Xtreme Everest Research Group: Angiotensin-converting enzyme genotype and successful ascent to extreme high altitude. High Alt Med Biol 2007; 8:278–285.

49 Tsianos G, Eleftheriou KI, Hawe E, Woolrich L, Watt M, Watt I, Peacock A, Montgomery H, Grant S: Performance at altitude and angiotensin I-converting enzyme genotype. Eur J Appl Physiol 2005; 93:630–633.

50 Costerousse O, Allegrini J, Lopez M, Alhenc-Gelas F: Angiotensin-I converting enzyme in human circulating mononuclear cells: genetic polymorphism of expression in T-lymphocytes. Biochem J 1993; 290:33–40.

51 Danser AHJ, Schalekamp MADH, Bax WA, van den Brink AM, Saxena PR, Reigger GAJ, Schunkert H: Angiotensin converting enzyme in the human heart: effect of the deletion/insertion polymorphism. Circulation 1995;92:1387–1388.

52 Davis GK, Millner RW, Roberts DH: Angiotensin converting enzyme (ACE) gene expression in the human left ventricle: effect of ACE gene insertion/deletion polymorphism and left ventricular function. Eur J Heart Fail 2000;2:253–256.

53 Mizuiri S, Hemmi H, Kumanomidou H, et al: Angiotensin-converting enzyme (ACE) I/D genotype and renal ACE gene expression. Kidney Int 2001;60:1124–1130.

54 Brown NJ, Blais C, Gandhi SK, Adam A: ACE insertion/deletion genotype affects bradykinin metabolism. J Cardiovasc Pharmacol 1998;32:373–377.

55 Ueda S, Elliott HL, Morton JJ, Connell JMC: Enhanced pressor response to angiotensin I in normotensive men with the deletion genotype (DD) for angiotensin-converting enzyme. Hypertension 1995;25:1266–1269.

56 Buikem H, Pinto YM, Rooks G, Grandjean JG, Schunkert H, van Gilst WH: The deletion polymorphism of the angiotensin-converting enzyme gene is related to phenotypic differences in human arteries. Eur Heart J 1996;17:787–794.

57 van Dijk MA, Kroon I, Kamper AM, Boomsma F, Danser AH, Chang PC: The angiotensin-converting enzyme gene polymorphism and responses to angiotensins and bradykinin in the human forearm. J Cardiovasc Pharmacol 2000;35:484–490.

58 Gordon SE, Davis BS, Carlson CJ, Booth FW: ANG II is required for optimal overload-induced skeletal muscle hypertrophy. Am J Physiol Endocrinol Metab 2001;280:E150–E159.

59 McBride TA: AT1 receptors are necessary for eccentric training-induced hypertrophy and strength gains in rat skeletal muscle. Exp Physiol 2006;91: 413–421.

60 Rattigan S, Dora KA, Tong AC, Clark MG: Perfused skeletal muscle contraction and metabolism improved by angiotensin II-mediated vasoconstriction. Am J Physiol 1996;271:E96–E103.

61 Brink M, Price SR, Chrast J, Bailey JL, Anwar A, Mitch WE, Delafontaine P: Angiotensin II induces skeletal muscle wasting through enhanced protein degradation and down-regulates autocrine insulin-like growth factor I. Endocrinology 2001;142:1489–1496.

62 Jones A, Woods DR: Skeletal muscle RAS and exercise performance. Int J Biochem Cell Biol 2003; 35:855–866.

63 Rett K, Wicklmayr M, Dietze GJ: Metabolic effects of kinins: historical and recent developments. J Cardiovasc Pharmacol 1990;15(suppl 6):S57–S59.

64 Shiuchi T, Nakagami H, Iwai M, Takeda Y, Cui T, Chen R, Minokoshi Y, Horiuchi M: Involvement of bradykinin and nitric oxide in leptin-mediated glucose uptake in skeletal muscle. Endocrinology 2001; 142:608–612.

65 Langberg H, Bjorn C, Boushel R, Hellsten Y, Kjaer M: Exercise-induced increase in interstitial bradykinin and adenosine concentrations in skeletal muscle and peritendinous tissue in humans. J Physiol 2002; 542:977–983.

66 Boix F, Røe C, Rosenborg L, Knardahl S: Kinin peptides in human trapezius muscle during sustained isometric contraction and their relation to pain. J Appl Physiol 2005;98:534–540.

67 Dietze GJ, Wicklmayr M, Rett K, Jacob S, Henriksen EJ: Potential role of bradykinin in forearm muscle metabolism in humans. Diabetes 1996;45:S110–S114.

68 Tanriverdi H, Evrengul H, Tanriverdi S, Turgut S, Akdag B, Kaftan HA, Semiz E: Improved endothelium dependent vasodilation in endurance athletes and its relation with ACE I/D polymorphism. Circ J 2005;69:1105–1110.

69 Butler R, Morris AD, Burchell B, Struthers AD: DD Angiotensin-converting enzyme gene polymorphism is associated with endothelial dysfunction in normal humans. Hypertension 1999;33:1164–1168.

70 Coyle EF, Sidossis LS, Horowitz JF, Beltz JD: Cycling efficiency is related to the percentage of type I muscle fibers. Med Sci Sports Exerc 1992;24:782–788.

71 Shimojo N, Chao J, Chao L, Margolius HS, Mayfield RK: Identification and characterization of a tissue kallikrein in rat skeletal muscles. Biochem J 1987; 243:773–778.

72 Ferrario CM, Iyer SN: Angiotensin-(1–7): a bioactive fragment of the renin-angiotensin system. Reg Pep 1998;78:13–18.

73 Rankinen T, Perusse L, Gagnon J, et al: Angiotensin-converting enzyme ID polymorphism and fitness phenotype in the HERITAGE Family Study. J Appl Physiol 2000;88:1029–1035.

74 Woods DR, World M, Rayson MP, Williams AG, Jubb M, Jamshidi Y, Hayward M, Mary DASG, Humphries SE, Montgomery HE: Endurance enhancement related to the human angiotensin I-converting enzyme I-D polymorphism is not due to differences in the cardiorespiratory response to training. Eur J Appl Physiol 2002;86:240–244.

75 Hagberg J, Ferrell M, McCole RE, Wilund SD, Moore GE: VO_{2max} is associated with ACE genotype in postmenopausal women. J Appl Physiol 1998; 85:1842–1846.

76 Williams AG, Rayson MP, Jubb M, World M, Woods DR, Hayward M, Martin J, Humphries SE, Montgomery HE: The ACE gene and muscle performance. Nature 2000;403:614.

77 Kanazawa H, Otsuka T, Hirata K, Yoshikawa J: Association between the angiotensin-converting enzyme gene polymorphisms and tissue oxygenation during exercise in patients with COPD. Chest 2002;21:697–701.

78 Zhang X, Wang C, Dai H, Lin Y, Zhang J: Association between angiotensin-converting enzyme gene polymorphisms and exercise performance in patients with COPD. Respirology 2008;13:683–688.

79 Day SH, Gohlke P, Dhamrait SS, Williams AG: No correlation between circulating ACE activity and VO_{2max} or mechanical efficiency in women. Eur J Appl Physiol 2007;99:11–18.

80 Zhang B, Tanaka H, Shono N, Miura S, Kiyonaga A, Shindo M, Saku K: The I allele of the angiotensin-converting enzyme gene is associated with an increased percentage of slow-twitch type I fibers in human skeletal muscle. Clin Genet 2003;63:139–144.

81 Dehnert C, Weymann J, Montgomery HE, Woods D, Maggiorini M, Scherrer U, Gibbs JS, Bartsch P: No association between high-altitude tolerance and the ACE I/D gene polymorphism. Med Sci Sports Exerc 2002;34:1928–1933.

82 Koehle MS, Wang P, Guenette JA, Rupert JL: No association between variants in the ACE and angiotensin II receptor 1 genes and acute mountain sickness in Nepalese pilgrims to the Janai Purnima Festival at 4380 m. High Alt Med Biol 2006;7:281–289.

83 Hotta J, Hanaoka M, Droma Y, Katsuyama Y, Ota M, Kobayashi T: Polymorphisms of renin-angiotensin system genes with high-altitude pulmonary edema in Japanese subjects. Chest 2004;126:825–830.

84 Woods DR, Pollard AJ, Collier DJ, Jamshidi Y, Vassiliou V, Hawe E, Humphries SE, Montgomery HE: Insertion/deletion polymorphism of the angiotensin I-converting enzyme gene and arterial oxygen saturation at high altitude. Am J Respir Crit Care Med 2002;166:362–366.

85 Bigham AW, Kiyamu M, León-Velarde F, Parra EJ, Rivera-Ch M, Shriver MD, Brutsaert TD: Angiotensin-converting enzyme genotype and arterial oxygen saturation at high altitude in Peruvian Quechua. High Alt Med Biol 2008;9:167–178.

86 Allen AM, Zhuo J, Mendelsohn FA: Localization and function of angiotensin AT1 receptors. Am J Hypertens 2000;13:31S–38S.

87 Patel S, Woods DR, Macleod NJ, Brown A, Patel KR, Montgomery HE, Peacock AJ: Angiotensin-converting enzyme genotype and the ventilatory response to exertional hypoxia. Eur Respir J 2003; 22:755–760.

88 Wang P, Fedoruk MN, Rupert JL: Keeping pace with ACE: are ACE inhibitors and angiotensin II type 1 receptor antagonists potential doping agents? Sports Med 2008;38:1065–1079.

Dr. David R. Woods
Department of Medicine, Royal Victoria Infirmary
Queen Victoria Road
Newcastle upon Tyne, NE1 4LP (UK)
Tel. +44 191 2336161, Fax +44 191 2010155, E-Mail DoctorDRWoods@aol.com

α-Actinin-3 and Performance

Nan Yang[a,b] · Fleur Garton[a,b] · Kathryn North[a,b]

[a]Institute for Neuroscience and Muscle Research, Children's Hospital at Westmead, and [b]Discipline of Paediatrics and Child Health, Faculty of Medicine, University of Sydney, Sydney, NSW, Australia

Abstract

The human sarcomeric α-actinins (*ACTN2* and *ACTN3*) are major structural components of the Z line in skeletal muscle; they play a role in the maintenance of sarcomeric integrity and also interact with a wide variety of structural, signaling and metabolic proteins. *ACTN2* is expressed in all muscle fibers, and expression of *ACTN3* is restricted to the type 2 (fast glycolytic) fibers that are responsible for forceful contraction at high velocity. There is a common stop codon polymorphism R577X in the *ACTN3* gene. Homozygosity for the R577X null-allele results in the absence of α-actinin-3 in fast muscle fibers with frequencies that vary from <1% in Africans to ~18% in Caucasians. A number of association studies have demonstrated that the *ACTN3* R577X genotype influences athletic performance in Caucasians; the frequency of the XX genotype is significantly lower than controls in sprint athletes, and it appears that α-actinin-3 deficiency is detrimental to sprint performance. In the general population, the *ACTN3* genotype contributes to the normal variations in muscle strength and sprinting speed. In an *Actn3* knockout mouse model, α-actinin-3 deficiency is associated with a shift in the characteristics of fast, glycolytic 2B muscle fibers towards a slow phenotype, with decreased muscle mass and fiber diameter, slower contractile properties, increased fatigue resistance, and an increase in oxidative enzyme activity. The shift towards a more efficient oxidative metabolism may underlie the selective advantage of the X-allele during evolution. In turn, the shift towards a 'slow' muscle phenotype in fast muscle fibers likely explains why loss of α-actinin-3 is detrimental to sprint performance.

Copyright © 2009 S. Karger AG, Basel

The Role of α-Actinins in Skeletal Muscle

The α-actinins are a family of actin-binding proteins that play a key role in the maintenance and regulation of the cytoskeleton [1]. In mammals, four α-actinin isoforms (α-actinin-1 to -4) have evolved from gene duplication events to fulfill similar functional requirements in different cell types [2–5]. *ACTN2* and *ACTN3* encode the skeletal muscle isoforms α-actinin-2 and α-actinin-3. *ACTN2* is expressed in all skeletal muscle fibers (both fast and slow fibers), as well as cardiac muscle and the brain. *ACTN3* expression is restricted to type 2 (fast glycolytic) fibers in skeletal muscle,

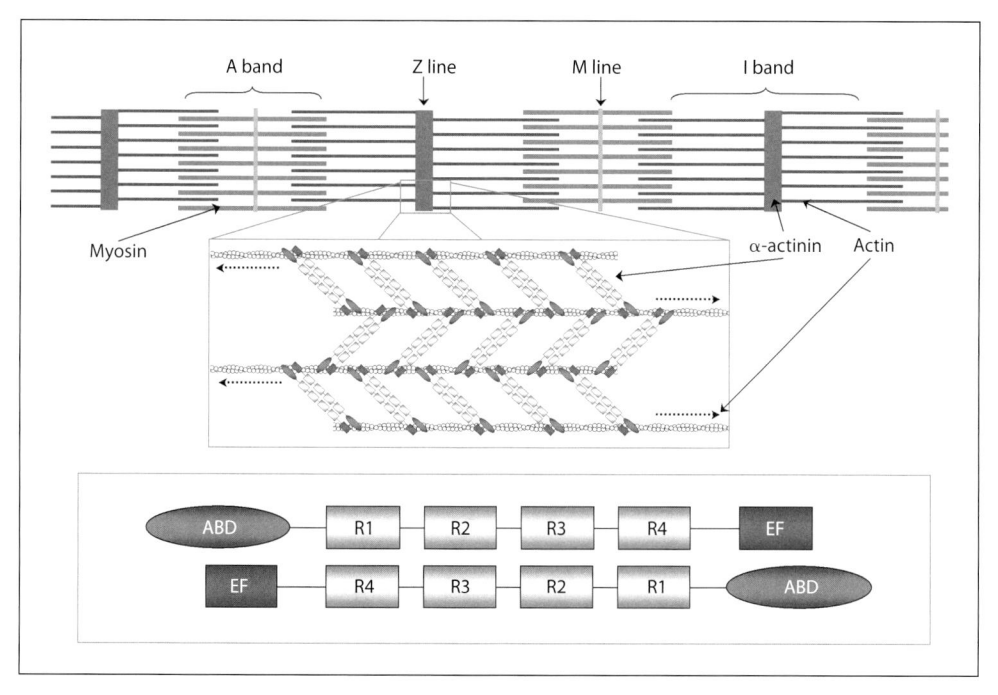

Fig. 1. Localization and domain structure of the sarcomeric α-actinins. The sarcomeric α-actinins are found at the Z line, where they anchor actin-containing thin filaments from adjacent sarcomeres. Antiparallel dimers of α-actinin cross-link actin filaments and stabilize them against force generated by the contractile apparatus (top inset; dashed arrows indicate direction of force). The lower inset illustrates the domain structure of the α-actinins and their antiparallel alignment in dimeric form. ABD = Actin-binding domain; R1–4 = spectrin-like repeats 1–4; EF = EF hand region. Adapted from MacArthur and North [44].

with low levels of expression in the brain [2]. The amino acid sequences of human *ACTN2* and *ACTN3* are 81% identical and 91% similar [2] and they form homodimers and heterodimers in vitro and in vivo, suggesting many interchangeable functions [6].

The contractile apparatus of skeletal muscle is composed of repeating units (sarcomeres) that contain ordered arrays of actin-containing thin filaments and myosin-containing thick filaments. The Z lines are electron-dense bands that run perpendicular to the myofibrils and anchor the thin filaments. The α-actinins are a major structural component of the Z line in skeletal muscle (fig. 1) and have a number of functional domains: an N-terminal actin-binding domain, a central rod domain, and a C-terminal region that contains two potential calcium-binding EF hand motifs [1]. The α-actinins provide a major multivalent platform for protein-protein interactions with many structural, signaling and metabolic proteins, resulting in diverse functional roles [7]. Based on their interaction with many of the structural and signaling proteins associated with the Z line, especially skeletal α-actin, the sarcomeric α-actinins (-2 and

-3) likely perform a static function in skeletal muscle to maintain the ordered myofibrillar array as well as a regulatory function in coordinating myofiber contraction [8].

The α-actinins interact with skeletal α-actin at the N-terminal and with other Z-line proteins (e.g. myotilin) at the C-terminal to organize and maintain the structure of the sarcomere in a three-dimensional lattice. The sarcomeric α-actinins are also thought to maintain muscle cell integrity by linking dystrophin (deficient in Duchenne muscular dystrophy) with the integrin adhesion system [9]. The interaction at the N-terminal with skeletal α-actin is regulated by binding of the phospholipid, phosphatidylinositol 4,5-biphosphate (PiP2) [10]. In vitro studies suggest that the C-terminus of sarcomeric α-actinins plays a crucial role in the maintenance of Z line integrity and myofibrillar organization [11]. The C-terminal region is the site of binding for a number of Z-line structural proteins, including myotilin [12] and a group of PDZ-LIM domain proteins such as ZASP and ALP [13]. Within the central repeat rod domain, the α-actinins interact with the Kv1.4 and Kv1.5 potassium channel proteins, fructose-1,6-bisphosphatase – a glycolytic enzyme [14], glycogen phosphorylase [15], and the calsarcins which bind to and regulate the expression of calcineurin – a signaling factor which plays a role in the specification of muscle fiber type, diameter and metabolism (oxidative versus glycolytic) [16]. Thus, differential expression of the sarcomeric α-actinins in different muscle fibers may influence their metabolic profile and/or fiber type specification (fig. 1).

α-Actinin-3 Deficiency Is Common in the General Population

The *ACTN3* gene contains a polymorphism, R577X, which results in conversion of the codon for arginine (R) at position 577 to a premature stop codon (X) [17]. This variation results in two versions of *ACTN3* in humans, a functional R-allele and a null X-allele. Homozygosity for the X-allele (XX genotype) results in complete deficiency of α-actinin-3 in humans [17]. The frequencies of the XX genotype differ in human populations ranging from ~1% in Eastern and Western Africans and South Africans, to ~18% in Europeans and up to ~25% in East Asians [18]. An estimated one billion people worldwide are deficient in α-actinin-3. The widespread deficiency of a structural protein which is highly restricted in expression in fast glycolytic fiber is very unusual, and raised the possibility that α-actinin-3 is functionally redundant, and that the closely related protein, α-actinin-2 is able to compensate for its loss in muscle. However, initial data suggested that *ACTN3* has function(s) independent of *ACTN2* in human and mouse muscles [18]. The experimental evidence can be summarized as follows:
1 *ACTN3* is highly conserved during evolution. Sequence comparison of human, mouse and chicken *ACTN* genes suggest that human *ACTN2* and *ACTN3* diverged substantially earlier than 300 million years ago, and the proteins encoded by both genes have evolved very slowly since their divergence.

2 *ACTN2* and *ACTN3* are differentially expressed, spatially and temporally, during human and mouse embryonic development, and *ACTN2* expression does not completely overlap *ACTN3* in mouse postnatal skeletal muscle. *ACTN3* is the predominant fast fiber isoform in both mice and humans. There are ~30% mouse 2B fibers with no *Actn*2 expression, whereas *Actn*3 is expressed in all 2B fibers.
3 The marked variation in *ACTN3* XX genotype frequencies in human populations raised the possibility that the *ACTN3* R577X genotype confers a fitness advantage to humans under certain environmental conditions.

Subsequent studies in human populations, in elite athletes and in an *Actn*3 KO mouse model have confirmed that *ACTN3* is not functionally redundant, that *ACTN2* cannot completely compensate for the loss of *ACTN3*, and that the presence or absence of α-actinin-3 influences human skeletal muscle performance. A detailed description of the experimental evidence is given in the following sections.

The Null X-Allele at the *ACTN3* Locus Has Undergone Recent Positive Natural Selection

The propagation of *ACTN3* deficiency in humans with marked variations in X-allele frequencies among different ethnic groups [17, 19] has raised the possibility that the X-allele has been positively selected in some populations. To explore the selective forces that have acted on the R577X polymorphism during human evolutionary history, Macarthur et al. [20] analyzed the DNA sequence base substitution rate and long-range linkage disequilibrium for recombination rate around R577X in individuals of European, East Asian and African ancestry. There was unusually low sequence diversity and high long-range linkage disequilibrium amongst X-allele-containing haplotypes compared with the R-allele in Europeans and Asians. This is consistent with strong, recent positive selection on X-allele in these populations – suggesting that the X-allele provided some sort of fitness advantage to modern humans adapting to the novel Eurasian environment [20].

***ACTN3* Genotype Is Associated with Elite Athletic Performance**

The force-generating capacity of glycolytic fast muscle fibers at high velocity and the capacity of the individual to adapt to exercise training are strongly genetically influenced, and contribute to the normal variation in muscle performance in the general population [21]. In addition, there appears to be an evolutionary predetermined 'trade-off' in performance traits for speed vs. endurance [22, 23] that is, an individual is inherently predisposed towards being more of a 'sprinter' or a 'stayer'.

α-Actinin-3 expression is restricted to fast muscle fibers and the R577X polymorphism has resulted in widespread α-actinin-3 deficiency in humans. This raised the

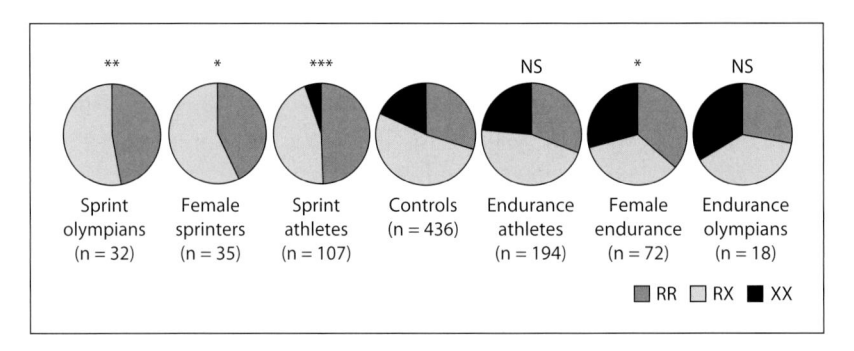

Fig. 2. Frequencies of the three *ACTN3* R577X genotypes in controls and elite athletes. The sample size in each group is shown. Female athletes and athletes who had competed at an Olympic level are shown as separate groups for both power/sprint and endurance athletes. Significant differences in genotype frequencies were seen between controls and total power athletes ($p < 0.0001$), female power athletes ($p < 0.05$), power Olympians ($p < 0.01$) and female endurance athletes ($p < 0.05$; χ^2 test for homogeneity of genotype frequencies between groups). Adapted from Yang et al. [24].

possibility that the *ACTN3* genotype contributes to normal human variation in skeletal muscle performance, and the presence or absence of α-actinin-3 may affect sprint or power activities performed by fast muscle fibers. Since any effect on muscle function will be most readily observable at the extremes of human performance, the effect of the *ACTN3* genotype has been most extensively studied in elite athletes.

Yang et al. [24] genotyped 301 elite Australian Caucasian athletes from 14 different sports and compared them with 436 healthy Caucasian controls. Athletes were defined as 'elite' if they had represented Australia in their sport at the international level; 50 of the athletes had competed at an Olympic level. There were considerable variations in *ACTN3* genotype frequencies between different sport groups of athletes, but there was no significant difference between genotype frequencies when comparing the athlete group as a whole with controls ($\chi^2 = 2.89$, $p = 0.06$). The R577X genotype was then analyzed in a subset of 107 specialist sprint/power athletes and compared with a subset of 194 specialist endurance athletes. There were 32 power athletes and 18 endurance athletes who had competed at the Olympic level. Remarkably, all the sprint/power athletes had similar genotypic profiles with an extremely low frequency of XX (5% compared with 18% in controls) and a high frequency of RR (50% compared with 30% in controls) ($\chi^2 = 19.70$, $p < 0.0001$). The genotype effect was more marked in females and in sprint/power Olympians (25 male and 7 female), none of whom were XX (fig. 2). In contrast, elite endurance athletes had higher frequencies of the XX genotype compared with controls, although the effect was significant only for females ($\chi^2 = 6.15$, $p < 0.05$). Importantly, genotype profiles in power and endurance athletes deviated in opposite directions and differed significantly from each other ($\chi^2 = 19.45$, $p < 0.001$), explaining the lack of allelic association when the entire elite athlete cohort was compared with the controls [24].

Following this study, there have been five independent association studies performed in elite sprint/power athletes which confirmed the initial finding that the *ACTN3* XX genotype occurs at significantly lower frequency in sprint/power athletes [25–29]. However, several association studies have failed to substantiate a significant association between α-actinin-3 deficiency and improved endurance performance – although there is a trend apparent in one study [24]. The results of these studies are summarized below and in tables 1–3.

The association studies of the *ACTN3* genotype in sprint/power athletes have included various sporting disciplines such as gymnastics, skating, swimming, track and field, throwing and weightlifting (table 1). All these typical mixed sprint/power athlete cohorts demonstrated significant association with the *ACTN3* genotype, with reduced XX and excess RR genotype frequencies compared with the relative control populations. The consistency of these results provides supportive evidence for a detrimental effect of the XX genotype and beneficial effect of the RR genotype on sprint/power performance.

The effect of the *ACTN3* genotype on athletic endurance performance is not clear. The Australian study [24] (total 193 athletes) and Russian study [30] (total 456 athletes) examined a wide range of endurance sports, including cycling, rowing, swimming, track and field and cross-country skiing (table 2). Australian elite male endurance athletes had a slightly higher XX genotype frequency (19%) and similar genotype distribution to Australian control (XX 16%), while Australian female endurance athletes had a significantly reduced RR (20%) and *increased* XX genotype (30%) frequencies compared with the control (RR 30% and XX 20%; table 2). However, in Russian endurance athletes both males and females had significantly *reduced* XX genotype frequency (4–7% in athletes compared with 13–17% in controls).

To clarify these contradictory results, we reviewed all published association studies of specific endurance sports – road cycling, ironman triathlon, long-distance running and rowing (table 3). Again the results show inconsistencies. In an Australian cohort of 75 road cyclists, there was a significant increase in the XX genotype (30%) compared with the control (XX 18%). In a similar Spanish road cycling cohort (n = 50), there was a slightly higher XX genotype frequency (26%) but it was not statistically different from the control (XX 18%) [42]. The other endurance-specific sports studied to date include a large cohort of participants in two ironman triathlons, long-distance runners competing in distances ranging from 3,000 m to marathon and rowers competing at distances greater than 2,000 m. All were found to have similar *ACTN3* XX genotype frequencies to ethnically similar controls (table 3). The combined weight of evidence from these studies suggests that the *ACTN3* XX genotype has no consistent beneficial or detrimental effect on endurance performance.

Studies in African athlete and control populations suggest that α-actinin-3 deficiency is not a major influence on performance in African athletes [19]. The frequency of the X allele was extremely low amongst Kenyans and Nigerians (~1% XX genotype) and higher in Ethiopians (~11% XX genotype). The low baseline frequencies

Table 1. *ACTN3* R577X genotypes in Sprint/Power athletic performance

Origin	Gender	Athletes n	sports	genotype, %	Controls n	genotype, %	p	Reference
Australian	M	72	judo; speed skating; swimming; track and field; track-cycling	RR: 53 RX: 39 XX: 8	134	RR: 30 RX: 54 XX: 16	<0.001	24
	F	35	judo; track and field; swimming	RR: 43 RX: 57 XX: 0	292	RR: 30 RX: 54 XX: 16	<0.01	
Finnish	M and F	23	track and field	RR: 48 RX: 52 XX: 0	120	RR: 45 RX: 46 XX: 9	0.03	25
Greek	M and F	73	sprint and power athletes (jumping; decathlon; throwing; and track and field)	RR: 48 RX: 36 XX: 16	181	RR: 26 RX: 56 XX: 18	<0.02	27
	M and F	34	sprinters (track 100–400 m)	RR: 73 RX: 18 XX: 9	181	RR: 26 RX: 56 XX: 18	<0.001	
Russian	M	363	alpine skiing; artistic gymnastics; bodybuilding; figure skating; ice hockey; power lifting; ski jumping; soccer; speed skating; swimming; throwing; track and field; volleyball; weightlifting; wrestling	RR: 38 RX: 56 XX: 6	524	RR: 37 RX: 47 XX:16	<0.0001	29
	F	123	alpine skiing; artistic gymnastics; bodybuilding; figure skating; ice hockey; power lifting; ski jumping; soccer; speed skating; swimming; throwing; track and field; volleyball; weightlifting; wrestling	RR: 46 RX: 48 XX: 6.5	673	RR: 37 RX: 51 XX: 13	0.067	
Spanish	M	60	soccer players	RR: 48 RX: 37 XX: 15	123	RR: 29 RX: 54 XX: 17	0.041	28
USA	M and F	75	bodybuilding; powerlifting; and college level strength	RR: 31 RX: 63 XX: 7	876	RR: 38 RX: 46 XX: 16	0.005	26

p value refers to the difference in XX genotype frequencies between athlete population and controls.

Table 2. ACTN3 R577X genotype in endurance athletic performance

Origin	Gender	Athletes n	sports	genotype, %	Controls n	genotype, %	p	Reference
Australian	M	118	cycling; rowing; swimming; track and field; XC skiing	RR: 28 RX: 53 XX: 19	134	RR: 30 RX: 54 XX: 16	NS	24
	F	75	cycling; rowing; swimming; track and field; XC skiing	RR: 20 RX: 50 XX: 30	292	RR: 30 RX: 50 XX: 20	<0.05	
Russian	M	293	biathlon ≥15 km; race walking ≥10 km; road cycling ≥50 km; rowing ≥2 km; swimming ≥0.8 km; triathlon; XC skiing ≥10 km	RR: 40 RX: 53 XX: 7	532	RR: 36 RX: 47 XX: 17	<0.01	30
	F	163	biathlon ≥15 km; race walking ≥10 km; road cycling ≥50 km; rowing ≥2 km; swimming ≥0.8 km; triathlon; XC skiing ≥10 km	RR: 37 RX: 59 XX: 4	679	RR: 37 RX: 50 XX: 13	<0.01	

p value refers to the difference in XX genotype frequencies between athlete population and controls.

of the three populations tested mean that any associations with sprint performance would likely be obscured. In Ethiopians, who excel in endurance sports, there was no increased frequency in the XX genotype in endurance athletes compared with controls.

Mechanisms Underlying the Effects of α-Actinin-3 Deficiency on Muscle Performance – Studies in *Actn3* KO Mouse and in Non-Athlete Humans

In order to understand the possible mechanisms underlying the effects of α-actinin-3 deficiency on muscle structural and physiological properties, we will now review studies in the α-actinin-3 knockout (*Actn3* KO) mouse model, and compare them with studies that have been performed to date in non-athlete human populations.

The *Actn3* KO mouse mimics human *ACTN3* deficiency in many ways [20]. The KO mice are morphologically indistinguishable from their WT littermates on gross physical inspection and general autopsy; they showed normal ultrastructure of Z line and actin thin filaments at the light and electron microscope level; they did not demonstrate substantial loss of fast (2B) fibers as defined by staining for myosin heavy extensor concentric peak torque (p = 0.019) and eccentric peak torque (p = 0.022) were lower in women with the XX genotype [34].

In a study of Greek adolescent boys (n = 525), XX individuals took significantly longer time to complete a 40-meter sprint (average time: XX = 6.13, RX = 6.00, RR = 5.92, p = 0.003) [35]. Interestingly, sprint times for individuals with the RR genotype were significantly faster than heterozygotes (RX), who were, in turn, significantly faster than XX – raising the possibility of a 'co-dominant' effect of the R and X allele – resulting in a 'dosage' effect relative to the amount of α-actinin-3 present in fast fibers. The potential heterozygote effect of the ACTN3 genotype has not yet been studied in any detail in either mice or humans.

α-Actinin-3 deficiency also appears to affect strength and power performance at the extremes of fitness. Within a group of human elite male road cyclists (n = 46), individuals with XX-genotypes were found to have less peak power output (p = 0.035), less power to tolerate high submaximal workloads (p = 0.029~0.01) compared with those with RR genotypes [36]. In a study to assess the genotype effect on ageing progression over a 5-year period, old women (70–79 y) with XX genotypes had a ~35% greater risk to develop difficulties in walking and climbing stairs compared with those with RR genotypes, and old men (70–79 years) with XX genotypes had greater increase in times to complete 400-meter walking [37]. Recently, the ACTN3 genotype has been shown to influence exercise capacity in women with the metabolic myopathy, McArdle's disease; those with the X allele had a higher VO_2 peak and a higher oxygen uptake (VO_2) at the ventilatory threshold than RR homozygotes [38].

Muscle Mass, Fiber Diameter and Fiber Type Proportion
Actn3 KO mice were shown to have a reduction in lean muscle mass using dual-energy X-ray absorptiometry scans [31]. This correlates with a decrease in the weight of individual muscles harvested from KO and WT mice. All muscles that contain fast-twitch fibers (and hence express α-actinin-3), including extensor digitorum longus, medial biceps, triceps, tibialis anterior, gastrocnemius, quadriceps and spinalis thoracis, have significant mass reduction in KO compared with WT [31]. Interestingly, a study in humans (n = 394) using a dual-energy X-ray absorptiometry scan also found a significant reduction in lean mass (p = 0.04) as well as fat mass (p = 0.02) in women with the XX genotype compared with women with combined RR and RX genotypes [34]. In this cohort, body mass index of women with the XX genotype was also significantly smaller than women with the RR genotype (p = 0.013) [34]. In a study examining the cross-sectional area (CSA) of human thigh muscle, postmenopausal women with the XX genotype had significantly reduced CSA compared with women with the RR genotype (p < 0.05) [39].

The reduction in muscle mass in *Actn3* KO mice is attributable to a specific decrease in fiber diameter of fast twitch (2B) muscle fibers – again reflecting a shift towards the phenotype of slow-twitch (type 1) fibers [31]. There was no difference in the diameter of type 1 or 2A fibers where α-actinin-3 is not usually expressed. There is tantalizing data – albeit limited by small study numbers – to suggest that the presence or absence of α-actinin-3 influences fiber diameter and fiber type in humans. In a cohort of young men (18–29 years), muscle biopsies were compared between 22 XX and 22 RR individuals. The percentage surface area and number of type 2X fibers (equivalent to 2B fibers in the mouse) were significantly less in those with the XX genotype compared with RR, while there were no significant differences in the size or number of type 1 and 2A fibers [33].

Conclusions

The *ACTN3* R577X polymorphism results in a well-defined phenotype (presence or absence of a muscle structural protein), and has a biologically plausible effect on skeletal muscle performance in athletes and in the general population; that is, loss of α-actinin-3 from fast-twitch muscle fibers is detrimental to sprint and power activities. The *Actn3* knockout mouse model provides mechanistic insights into the effects of the *ACTN3* genotype in humans. Loss of α-actinin-3 in mouse results in a shift in a variety of properties of fast fibers towards those characteristic of slower fibers. The slower contractile properties, reduced fiber diameter and decrease in anaerobic metabolism associated with α-actinin-3 deficiency explains the detrimental effect on sprint and power performance observed in multiple Caucasian athlete association studies. The shift in muscle metabolism toward the more efficient aerobic pathway may have provided adaptive benefits during a period of human evolution when man migrated out of Africa and faced famine and cold; perhaps explaining the strong positive selection for the X allele in Europeans and Asians. The X allele is much rarer in African populations (XX genotype frequencies in most African populations are less than 1%). Thus, the *ACTN3* genotype is unlikely to be an important determinant of variations in human performance in African populations. Whether or not α-actinin-3 deficiency influences adaptations for endurance performance is less clear; there may be more subtle effects that remain to be explored such as the effect on response to exercise and training.

Variations in the *ACTN3* R577X genotype may also have public health implications; an estimated one billion people worldwide are deficient in α-actinin-3. The reduced muscle mass and strength associated with α-actinin-3 deficiency in the general population is within the range of normal human variation. However, α-actinin-3 deficiency could modify quality of life and severity of muscle weakness at the extremes of human fitness – for example, in the less mobile ageing population prone to muscle sarcopenia, in patients with cachexia due to chronic illness and in patients with muscle disease. Intriguingly, the alterations in muscle metabolism associated with the *ACTN3* genotype could modify response to diet and the risk of developing common disorders in the modern world such as type 2 diabetes and obesity. Future studies of the advantages and disadvantages that the ACTN3 X-allele has conferred on humans will provide interesting insights into how genetic factors that influenced survival in ancient humans contribute to performance and health in modern man.

References

1 Blanchard A, Ohanian V, Critchley D: The structure and function of alpha-actinin. J Muscle Res Cell Motil 1989;10:280–289.
2 Beggs AH, Byers TJ, Knoll JH, Boyce FM, Bruns GA, Kunkel LM: Cloning and characterization of two human skeletal muscle alpha-actinin genes located on chromosomes 1 and 11. J Biol Chem 1992;267:9281–9288.
3 Honda K, Yamada T, Endo R, et al: Actinin-4, a novel actin-bundling protein associated with cell motility and cancer invasion. J Cell Biol 1998;140: 1383–1393.
4 Southby J, Gooding C, Smith CW: Polypyrimidine tract binding protein functions as a repressor to regulate alternative splicing of alpha-actinin mutually exclusive exons. Mol Cell Biol 1999;19:2699–2711.
5 Kremerskothen J, Teber I, Wendholt D, Liedtke T, Bockers TM, Barnekow A: Brain-specific splicing of alpha-actinin 1 (ACTN) mRNA. Biochem Biophys Res Commun 2002;295:678–681.
6 Chan Y, Tong HQ, Beggs AH, Kunkel LM: Human skeletal muscle-specific alpha-actinin-2 and -3 isoforms form homodimers and heterodimers in vitro and in vivo. Biochem Biophys Res Commun 1998; 248:134–139.
7 Djinovic-Carugo K, Gautel M, Ylanne J, Young P: The spectrin repeat: a structural platform for cytoskeletal protein assemblies. FEBS Lett 2002;513: 119–123.
8 MacArthur DG, North KN: A gene for speed? The evolution and function of alpha-actinin-3. Bioessays 2004;26:786–795.
9 Hance JE, Fu SY, Watkins SC, Beggs AH, Michalak M: Alpha-actinin-2 is a new component of the dystrophin-glycoprotein complex. Arch Biochem Biophys 1999;365:216–222.
10 Yin HL, Janmey PA: Phosphoinositide regulation of the actin cytoskeleton. Annu Rev Physiol 2003;65: 761–789.
11 Lin Z, Hijikata T, Zhang Z, et al: Dispensability of the actin-binding site and spectrin repeats for targeting sarcomeric alpha-actinin into maturing Z bands in vivo: implications for in vitro binding studies. Dev Biol 1998;199:291–308.
12 Salmikangas P, van der Ven PF, Lalowski M, et al: Myotilin, the limb-girdle muscular dystrophy 1A (LGMD1A) protein, cross-links actin filaments and controls sarcomere assembly. Hum Mol Genet 2003;12:189–203.
13 Sjoblom B, Salmazo A, Djinovic-Carugo K: Alpha-actinin structure and regulation. Cell Mol Life Sci 2008;65:2688–2701.
14 Gizak A, Rakus D, Dzugaj A: Immunohistochemical localization of human fructose-1,6-bisphosphatase in subcellular structures of myocytes. Histol Histopathol 2003;18:135–142.
15 Chowrashi P, Mittal B, Sanger JM, Sanger JW: Amorphin is phosphorylase; phosphorylase is an alpha-actinin-binding protein. Cell Motil Cytoskeleton 2002;53:125–135.
16 Semsarian C, Wu MJ, Ju YK, et al: Skeletal muscle hypertrophy is mediated by a Ca^{2+}-dependent calcineurin signalling pathway. Nature 1999;400:576–581.
17 North KN, Yang N, Wattanasirichaigoon D, Mills M, Easteal S, Beggs AH: A common nonsense mutation results in alpha-actinin-3 deficiency in the general population. Nat Genet 1999;21:353–354.
18 Mills M, Yang N, Weinberger R, et al: Differential expression of the actin-binding proteins, alpha-actinin-2 and -3, in different species: implications for the evolution of functional redundancy. Hum Mol Genet 2001;10:1335–1346.
19 Yang N, MacArthur DG, Wolde B, et al: The ACTN3 R577X polymorphism in East and West African athletes. Med Sci Sports Exerc 2007;39:1985–1988.
20 MacArthur DG, Seto JT, Raftery JM, et al: Loss of ACTN3 gene function alters mouse muscle metabolism and shows evidence of positive selection in humans. Nat Genet 2007;39:1261–1265.
21 Simoneau JA, Bouchard C: Variation on Anaerobic performance; in Praagh E (ed): Pediatric Anaerobic Performance. Champaign, Human Kinetics, 1998, pp 5–19.
22 Van Damme R, Wilson RS, Vanhooydonck B, Aerts P: Performance constraints in decathletes. Nature 2002;415:755–756.
23 Garland T, Bennett AF, Daniels CB: Heritability of locomotor performance and its correlates in a natural population. Cell Mol Life Sci 1990;46:530–533.
24 Yang N, MacArthur DG, Gulbin JP, et al: ACTN3 genotype is associated with human elite athletic performance. Am J Hum Genet 2003;73:627–631.
25 Niemi AK, Majamaa K: Mitochondrial DNA and ACTN3 genotypes in Finnish elite endurance and sprint athletes. Eur J Hum Genet 2005;13:965–959.

26 Roth SM, Walsh S, Liu D, Metter EJ, Ferrucci L, Hurley BF: The ACTN3 R577X nonsense allele is under-represented in elite-level strength athletes. Eur J Hum Genet 2008;16:391–394.
27 Papadimitriou ID, Papadopoulos C, Kouvatsi A, Triantaphyllidis C: The ACTN3 gene in elite Greek track and field athletes. Int J Sports Med 2008;29:352–355.
28 Santiago C, Gonzalez-Freire M, Serratosa L, et al: ACTN3 genotype in professional soccer players. Br J Sports Med 2008;42:71–73.
29 Druzhevskaya AM, Ahmetov, II, Astratenkova IV, Rogozkin VA: Association of the ACTN3 R577X polymorphism with power athlete status in Russians. Eur J Appl Physiol 2008;103:631–634.
30 Ahmetov, II, Druzhevskaya AM, Astratenkova IV, Popov DV, Vinogradova OL, Rogozkin VA: The ACTN3 R577X polymorphism in Russian endurance athletes. Br J Sports Med 2008, Epub ahead of print.
31 MacArthur DG, Seto JT, Chan S, et al: An Actn3 knockout mouse provides mechanistic insights into the association between alpha-actinin-3 deficiency and human athletic performance. Hum Mol Genet 2008;17:1076–1086.
32 Clarkson PM, Devaney JM, Gordish-Dressman H, et al: ACTN3 genotype is associated with increases in muscle strength in response to resistance training in women. J Appl Physiol 2005;99:154–163.
33 Vincent B, De Bock K, Ramaekers M, et al: ACTN3 (R577X) genotype is associated with fiber type distribution. Physiol Genomics 2007;32:58–63.
34 Walsh S, Liu D, Metter EJ, Ferrucci L, Roth SM: ACTN3 genotype is associated with muscle phenotypes in women across the adult age span. J Appl Physiol 2008;105:1486–1491.
35 Moran CN, Yang N, Bailey ME, et al: Association analysis of the ACTN3 R577X polymorphism and complex quantitative body composition and performance phenotypes in adolescent Greeks. Eur J Hum Genet 2007;15:88–93.
36 Gomez-Gallego F, Santiago C, Gonzalez-Freire M, et al: Endurance performance: genes or gene combinations? Int J Sports Med 2009;30:66–72.
37 Delmonico MJ, Zmuda JM, Taylor BC, et al: Association of the ACTN3 genotype and physical functioning with age in older adults. J Gerontol A Biol Sci Med Sci 2008;63:1227–1234.
38 Lucia A, Gomez-Gallego F, Santiago C, et al: The 577X allele of the ACTN3 gene is associated with improved exercise capacity in women with McArdle's disease. Neuromuscul Disord 2007;17:603–610.
39 Zempo H, Tanabe K, Murakami H, Iemitsu M, Maeda S, Kuno S: Association of muscle related factor polymorphisms and thigh muscle cross sectional area in Japanese postmenpausal women; in 13th Annu Eur Coll Sport Sci, Estoril, 2008.
40 Paparini A, Ripani M, Giordano GD, Santoni D, Pigozzi F, Romano-Spica V: ACTN3 genotyping by real-time PCR in the Italian population and athletes. Med Sci Sports Exerc 2007;39:810–815.
41 Saunders CJ, September AV, Xenophontos SL, Cariolou MA, Anastassiades LC, Noakes TD, Collins M: No association of the ACTN3 gene R577X polymorphism with endurance performance in Ironman Triathlons. Ann Hum Genet 2007;71:777–781.
42 Lucia A, Gómez-Gallego F, Santiago C, Bandrés F, Earnest C, Rabadán M, Alonso JM, Hoyos J, Córdova A, Villa G, Foster C: ACTN3 genotype in professional endurance cyclists. Int J Sports Med 2006;27:880–884.
43 Muniesa CA, González-Freire M, Santiago C, Lao JI, Buxens A, Rubio JC, Martín MA, Arenas J, Gomez-Gallego F, Lucia A: World-class performance in lightweight rowing: is it genetically influenced? A comparison with cyclists, runners and non-athletes. Br J Sports Med, in press.
44 MacArthur D, North KN: A gene for speed? The function and evolutionary history of 4-actinin-3. Bioessays 2004;26:786–895.

Prof. Kathryn North
Institute for Neuroscience and Muscle Research
Children's Hospital at Westmead, Locked Bag 4001
Westmead, NSW 2145 (Australia)
Tel. +61 2 9845 1906, Fax +61 2 9845 3389, E-Mail kathryn@chw.edu.au

East African Runners: Their Genetics, Lifestyle and Athletic Prowess

Vincent O. Onywera

Department of Exercise, Recreation and Sports Science, and IAAF Athletics Academy at Kenyatta University, Nairobi, Kenya

Abstract

East African runners have dominated distance running events for over 5 decades. Some explanations have been advanced to explain why such a small population has dominated distance running events over time. Suggested reasons include, among others, a genetic predisposition, diet, living at high altitude as well as sociocultural background. This chapter gives possible insight into the past, present and hopefully future success of East African runners; it mainly explores the foundations of running excellence, talent identification, diet and injury management methods used by East African runners. The chapter also explores means and ways by which East African runners can sustain their running excellence by using their past experiences, to perfect the present and predict the future.

Copyright © 2009 S. Karger AG, Basel

The continued excellence of East African athletes in distance running events continues to produce a lot of debate and interest across the globe. In fact, there has been an overabundance of opinions and theories aimed at explaining the phenomenal athletic prowess of the 'East African running machines – the exquisite runners'. Authors have suggested a superior genetic predisposition, sociocultural upbringing and sound nutrition as some of the possible factors behind the success of East African runners [1–7]. Runners from the eastern part of Africa nevertheless continue to enjoy unrivalled success in distances ranging from 800 meters to the marathon [8]. This chapter gives an academic discourse of the role that possible candidate genes coupled with habitual lifestyle could have lead to the success of East African runners.

Genetics of Performance and the East African Runners

The genetic predisposition (nature) of *Homo sapiens* plays an important role in determining many characteristics including ability to succeed in athletics and sports. Our diet, upbringing and habitual physical activity also play a large role in enhancing and developing our athletic potential. One may, however, have the genetic potential to become a world-class athlete, but if one has not been nurtured to develop athletic excellence then one is unlikely to achieve that potential [9]. It is, however, worth noting that our genes may also limit our performance [9, 10].

Genetics have a large influence over many attributes necessary for athletic excellence such as strength, muscle size and muscle fiber composition (fast or slow twitch), anaerobic threshold, lung capacity, flexibility [11]. An athlete's cardiac capacity, or the heart's ability to deliver enough oxygen to the working skeletal muscles, for example, is a major determinant in endurance athletics. This too, is largely determined by genetics.

Research on aerobic endurance has shown that some people respond more to training than others [12]. Therefore, even if one has a low genetic potential for endurance, one may respond well to training and develop one's potential more completely than someone with a genetic 'talent' who does not respond to training. Training also increases cardiac efficiency, but the extent of this increase may be determined genetically [12].

It is hypothesized that great runners such as those from the eastern part of Africa are endowed with 'running' genes [7]. Several studies have been conducted to date to elucidate the role of genes in sports performance. Some studies have for example showed a significantly lower frequency of the 577XX (α-actinin-3-deficient) genotype in sprint or power athletes compared with controls, suggesting that the presence of wild-type α-actinin-3 enhances performance in sprint-type activities and an increase in the frequency of the 577XX genotype has been documented in some, but not all, endurance athletes [13, 14].

The role of the ACE gene, one of the most widely studied genes in recent times, has also been implicated in athletic performance [10, 15]. A study by Collins et al., [16] found a significantly higher frequency of the I-allele of the ACE gene within the fastest 100 South African-born finishers compared with the 166 South African-born control subjects. Myerson et al. [17] found an increasing frequency of the I-allele with distance running. A study investigating the association of DNA sequence variants within the ACE, angiotensinogen and angiotensin receptor type 1 genes with 60 elite (professional) athletes (25 cyclists, 20 long-distance runners, and 15 handball players) and 400 healthy controls showed that the I-allele occurred at a significantly higher frequency in athletes compared with controls [18].

Scott et al. [19] assessed the association between ACE gene variations and elite endurance athlete status in an African population successful in distance running. DNA samples were obtained from 221 national Kenyan athletes, 70 international Kenyan athletes, and 85 members of the general Kenyan population. Previous findings that many of the most successful East African athletes reside and train at altitude [4, 20] may have supported a role for the ACE gene in the determination of their success in distance running. Many Kenyan athletes live and train at altitudes of over 2,500 m

[6]. It is therefore conceivable that there has been selection for the I-allele of the ACE gene amongst such a population which may contribute to the altitude tolerance and endurance performance phenotypes. Results from the study by Scott et al. [19] did not support this hypothesis as there was an absence of an association between either I/D or A22982G polymorphisms within the ACE gene and elite Kenyan athlete status. For the East African group, the ACE gene therefore probably does not contribute significantly to the phenomenal success of Kenyan endurance runners in international distance running competition.

In another study, Yang et al. [21] looked at the ACTN3 R577X polymorphism among East and West African athletes. The R577X polymorphism was genotyped in 198 Ethiopian controls and 76 elite Ethiopian endurance athletes, 158 Kenyan controls and 284 elite Kenyan endurance runners, and 60 Nigerian controls and 62 elite Nigerian power athletes. The study found no significant deviations from the Hardy-Weinberg equilibrium in the populations tested. The XX (α-actinin-3 null) genotype was absent in Nigerians, had an extremely low frequency among Kenyans (about 1%), and a higher frequency in Ethiopians (about 11%). The study, therefore, did not support the hypothesis that the ACTN3 null X-allele contributes to the success of the East African endurance runners relative to their source populations. In addition, the extremely low XX genotype frequency in the Nigerian control population (0% in this study) ruled out the possibility of detecting a lower frequency of the XX genotype in Nigerian power athletes.

In a related study, Scott et al. [22] compared the frequencies of mtDNA haplogroups found in elite Kenyan endurance athletes with those in the general Kenyan population. DNA samples were obtained from 221 national level Kenyan athletes (N), 70 international Kenyan athletes (I), and 85 members of the general Kenyan population (C). mtDNA was classified into haplogroups by sequencing 340 bases of hypervariable sequence (HVS I), and by genotyping known restriction sites. Frequency differences between groups were assessed using exact tests of population differentiation. Results showed that the haplogroup distribution of national ($p = 0.023$) and international athletes ($p < 0.001$) differed significantly from controls, with the international athletes showing a greater proportion of L0 haplogroups (C = 15%, N = 18%, I = 30%), and lower proportion of L3* haplogroups (C = 48%, N = 36%, I = 26%). Although many of international athletes originated from the Rift Valley province in Kenya relative to controls (C = 20%, N = 65%, I = 81%), this province did not differ in haplogroup distribution from other regions ($p = 0.23$). Similarly, although Nilotic ethnicities were overrepresented amongst the athletes (C = 21%, N = 60%, I = 79%), haplogroup distributions did not differ between Bantu and Nilotic subjects ($p = 0.12$). The study concluded that International athletes differed in their mtDNA haplogroup distribution relative to the general Kenyan population. They displayed an excess of L0 haplogroups, and a dearth of L3* haplogroups. These findings are consistent with the idea that certain mtDNA haplogroups are influential in elite Kenyan distance running, although population stratification cannot be ruled out.

The Inspiration of Young Runners from East Africa

There are a number of reasons that can account for the motivation and inspiration of young and upcoming runners from East Africa [8]. These include and are not limited to:
1 Exposure to running at a young age,
2 Contact with training groups, which include international athletes, in local areas,
3 The presence and freedom to mix and mingle with role models,
4 Opportunities to train alongside an Olympian or World Champion,
5 Athletics is viewed as a 'gateway to fame and riches'.

The Diet and Way of Life of East African Runners

Nutritional supplements are still largely unknown among most East African athletes. The staple diet of *'ugali'*, accompanied by traditional vegetables *'isochot'*, *'osuga'* and *'isakiat'* are considered good for energy. *Ugali* also sometimes called *sima* or *sembe* is a cornmeal product and a staple starch component of many African meals, especially in East Africa. It is generally made from maize flour (or ground maize) and water, and varies in consistency from porridge to a dough-like substance.

The drinking of milk and blood is still common among some communities in Kenya, especially the Kalenjin and Masai. Breakfast consists of *'uji'* – a porridge made of millet. There is for example no 'junk food', no chocolate, and no ice cream in Kenyan villages. Children living in these villages are also not aware of these foods. A study by Onywera et al. [20] on the dietary habits of 12 elite Kenyan runners for a 7-day period found that the estimated energy intake (EI) over the 7-day assessment period (2,987 ± 293 kcal) was significantly lower than estimated (using PAR) energy expenditure (EE; 3,605 ± 119 kcal; $p < 0.001$). This is also reflected by the low EI:EE ratio (0.85 ± 0.08). The average daily EI of the runners was approximately 620 kcal (17%) below the average estimated daily EE for the athletes.

Most young athletes in East Africa run barefooted. In fact, shoes are still a novelty in most parts of Africa. We cannot forget Christopher Kosgei during the Grand Prix meets in Europe and Peter Chumba who won two gold medals during the first World Juniors Championships in Athens in 1986 bare feet. The entire junior women's gold medal winning team at the 1993 World Cross Country Championships in Spain competed bare foot. It is believed that direct contact with the hard ground strengthens the feet of the runners right from an early age. Although injuries are inevitable in any sporting event, East Africans have their own traditional way of managing them. Some of the traditional remedies for injuries include:
1 Bathing wounds in salty hot water,
2 Applying the cream of the milk to remove a thorn,
3 The Kalenjin tribe rubbing the leaves of the *'irokwet'* tree to a bruise.

The only cure that is used by many runners especially in the rural areas for stiff muscles or soreness is rest or rubbing Vaseline, animal fat, the bark or leaves of certain trees on the stiff or sore muscles.

Athletics Facilities before 'Westernization'

During the time preceding the 1980s, few schools in Kenya could afford their own track and none had a gymnasium or indoor facility. Athletes trained on the dirt tracks and pathways. They also used crude and/or improvised equipment [8, 23]. Today, many still refer to the javelin as a 'spear' and the shot as 'the stone' – a reference to what was used several years ago. Stopwatches were a novelty for many until recently. Few athletes knew their times or splits. We are fairly sure that there are few or no stadium records in Kenya. The idea of electronic timing is still a dream except, perhaps, at the national senior championships or trials. For the purpose of weight training, specific exercises and monitoring equipment, schools like St. Patrick's High School located in Iten at 8,000 feet above sea level, had to do with improvisation and guesswork. It is worth noting that St. Patrick's High School has, over the last 30 years, produced world-class long distance athletes. Some of the alumni include Ibrahim Hussein, winner of three Boston Marathons and one New York City Marathon; Peter Rono, a 1988 Olympic gold medalist at 1,500 m; Wilson Boit Kipketer, a 1997 world champion and 2000 Olympic silver medalist in the 3,000-meter steeplechase. Athletes used 'homemade' weights with little reference to accuracy or safety. Heart rate monitors, treadmills, let alone GPS locators, are items that are only read about in books [8].

Advantages of Such a Situation

The above scenario had some advantages. It prevented coaches and trainers from setting limits to workouts and programs, from measuring and waiting to be told by a machine what you already knew through common sense and instinct. Do you need an instrument to tell you that you're training too hard or that you're tired? Or that you have an injury to a knee? Learning to listen to their bodies and how the athletes feel was and still is an important part of a monitoring and evaluating process in Kenya. St. Patrick's for example has no track or gymnasium. They depend on a dilapidated stadium about 4 km away from the school, with a poorly maintained dirt track which becomes muddy during the rainy season. They have to make full use of the open countryside, its quietness and isolation, the closeness of nature, the dirt roads, the hills and forests.

Talent Identification and Nurturing

Most coaches often stand on the sideline during a primary schools athletics competition and wonder if someone out there has the potential to be an Olympic champion. They identify talented runners based on the 'theory of probability'. This takes into account the interest, personality, background, education, body structure, posture, movement, running style, pace, body size and shape, muscular structure and alignment of a particular runner. Br. Colm says: 'Having had little or not technical background to coaching athletics, I depended very much upon my ability to observe and learn from the athlete. There was an element of "trial and error" for me and, to some extent, still is.' [Cornnel, pers. commun., 2004].

The Role of Parents in Athletics Nurturing in East Africa

For many years, parents played little or no role in athletic development of their children. Many parents still consider running as a pastime or entertainment for youngsters rather than a profession. Some may feel that their children could have 'better things to do'. We encourage parents to be part of their children's athletic development. This is especially true of 'the girl athlete'. Traditionally girls were married off for dowry, they had their role in the home and their place was in the kitchen. An athletic career had little to do with it. When identifying a girl athlete – we ensure that parents and family understand the implications and give support to their daughter. Married women in Kenya were almost banned from competitions in the 1960s and 1970s. It was taboo for a married lady to dress in shorts and vest with bare legs and arms. Although girls had represented Kenya at major competitions, there is not yet a female 'running culture'.

The Present Scenario with Regard to Training and Preparation of Athletes

Things have not changed much especially with regard to training and preparation of athletes for local and international competitions. Many athletes still 'listen' to their bodies during training and competitions. Most of the athletic coaches are either former athletes themselves or officers working for the armed forces or police. Crude methods of training and monitoring of performance are still very popular.

Kenyan athletes have stuck to the past ways of living and training. They still cherish the lives lead by the likes of Mike Boit, Abebe Bikila, Kipchoge Keino and Amos Biwott among other pioneer runners. Most African athletes and coaches have not appreciated the role and place of science in boosting their performance and excellence on the track and in the field. As illustrated in figure 1, Onywera et al. [6] noted that Kenyan distance runners are motivated by economic reasons

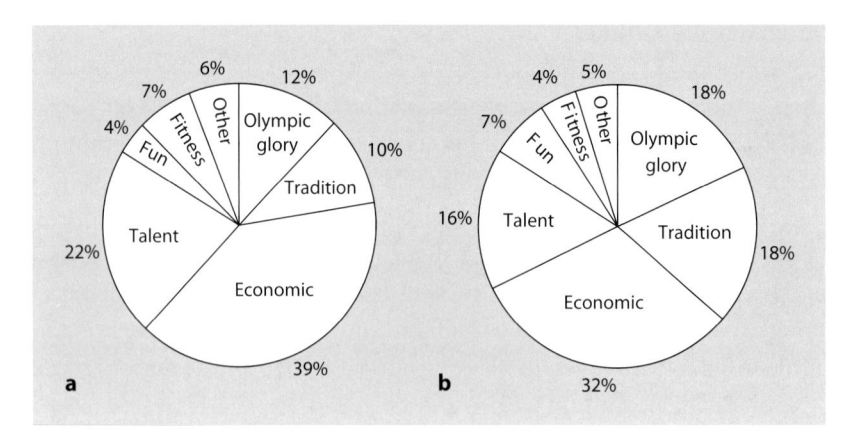

Fig. 1. Motivation to become a competitive athlete. The national (**a**) and international (**b**) athlete groups differed significantly from each other.

although 18% of international athletes, especially those who ran in the 1970s and 1980s, ran for Olympic glory.

Poverty and unemployment are major cancers in most African countries. It would be understandable, therefore, that economic reasons act as an important contributing factor to the success of East African athletes in distance running, a factor not accounted for in any other studies to date.

Conclusion

Having looked at various studies on the genetics of human performance as well as the way of life of East African runners, it is clear that more research needs to be done to unravel the mystery that is East African running excellence. More studies and resources are needed to explain the phenomenal success of East African runners.

References

1 Saltin B: Possible Explanations for the Success of Kenyan Middle and Long Distance Runners. New Studies in Athletics. Monaco, IAAF, 2003.
2 Saltin B, Larsen H, Terrados N, Bangsbo J, Bak T, Kim CK, Svedenhag J, Rolf CJ: Aerobic exercise capacity at sea level and at altitude in Kenyan boys, junior and senior runners compared with Scandinavian runners. Scand J Med Sci Sports 1995; 5:209–221.
3 Saltin B: Exercise and the environment: focus on altitude. Res Q Exerc Sport 1996;67(suppl 3):S1–S10.
4 Scott RA, Georgiades E, Wilson RH, Goodwin WH, Wolde B, Pitsiladis YP: Demographic characteristics of elite Ethiopian endurance runners. Med Sci Sports Exerc 2003;35:1727–1732.
5 Pitsiladis YP, Onywera VO, Geogiades E, O'Connell W, Boit MK: The dominance of Kenyans in distance running. Equine Compar Exerc Physiol 2004;1:285–291.
6 Onywera VO, Scott RA, Boit MK, Pitsiladis YP: Demographic characteristics of elite Kenyan endurance runners. J Sports Sci 2006;24:415–422.

7 Scott R, Pitsiladis YP: Genetics and the success of East African distance runners. Int Sportmed J 2006;7:172–186.
8 Boit MK: Kenyan Runners. In Search of Olympic Glory. Nairobi, Jomo Kenyatta Foundation, 2004.
9 Bouchard C, Malina R, Perusse L: Genetics of Fitness and Physical Performance. Champaign, Human Kinetics, 1997, pp 1–400.
10 Montgomery HE, Marshall R, Hemingway H, Myerson S, Clarkson P, Dollery C, et al: Human gene for physical performance. Nature 1998;393: 221–222.
11 di Prampero PE: Factors limiting maximal performance in humans. Eur J Appl Physiol 2003;90:420–429.
12 Skinner JS, Jaskolski A, Jaskolska A, Kransnoff J, Gagon J, Leon AS, Rao DC, Wilmore JH, Bouchard C: Age, sex, race, initial fitness, and response to training: The HERITAGE Family Study. J Appl Physiol 2001;90:1770–1776.
13 MacArthur DG, North KN: Bioessays. 2004;26:786–795.
14 Yang N, MacArthur DG, Gulbin JP, Hahn AG, Beggs AH, Easteal S, et al: ACTN3 genotype is associated with human elite athletic performance. Am J Hum Genet 2003;73:627–631.
15 Woods D, Hickman M, Jamshidi Y, Brull D, Vassiliou V, Jones A, et al: Elite swimmers and the D allele of the ACE I/D polymorphism. Hum Genet 2001; 108:230–232.
16 Collins M, et al: The ACE gene and endurance performance during the South African Ironman Triathlons. Med Sci Sports Exerc 2004;36:1314–1320.
17 Myerson S, et al: Human angiotensin I-converting enzyme gene and endurance performance. J Appl Physiol 1999;87:1313–1316.
18 Alvarez R, et al: Genetic variation in the renin-angiotensin system and athletic performance. Eur J Appl Physiol 2000;82:117–120.
19 Scott R, Moran CN, Wilson RH, Onywera VO, Boit M, Goodwin WH, Gohlke P, Montgomery H, Pitsiladis YP: No association between angiotensin converting enzyme (ACE) gene variation and endurance athlete status in Kenyans. Comp Biochem Physiol A Mol Integr Physiol 2005;141:169–175.
20 Onywera VO, Kiplamai FK, Tuitoek, PJ, Boit MK, Pitsladis YP: Food and macronutrient intake of elite Kenyan distance runners. Int J Sports Nutr 2004; 14:709–719.
21 Yang N, MacArthur DG, Wolde B, Onywera VO, Boit MK, Mary-Ann Lau SY, Wilson RH, Scott RA, Pitsiladis YP, North K: The ACTN3 R577X polymorphism in East and West African athletes. Med Sci Sports Exerc 2007;39:1985–1988.
22 Scott R, Fuku N, Onywera VO, Wilson RH, Boit M, Goodwin WH, Tanaka M, Pitsiladis YP: Mitochondrial haplogroups associated with elite Kenyan athlete status. Med Sci Sports Exerc 2009;41:123–128.
23 Bale J, Sang J: Kenyan Running. Movement, Culture, Geography and Global Change. London, Frank Cass, 1996.

Vincent O. Onywera, PhD
Kenyatta University
Department of Exercise, Recreation and Sport Science
PO BOX 43844-00100 Nairobi (Kenya)
Tel. +254 2 810 901, ext. 57 284, Fax +254 20 811 575
E-Mail vonywera@yahoo.com, onywera.vincent@ku.ac.ke

Gene-Lifestyle Interactions and Their Consequences on Human Health

Jeremy Pomeroy[a,b] · Anna M. Söderberg[b] · Paul W. Franks[b]

[a]Phoenix Epidemiology and Clinical Research Branch, National Institute of Diabetes and Digestive and Kidney Diseases, National Institutes of Health, Phoenix, Ariz., USA; [b]Genetic Epidemiology and Clinical Research Group, Department of Public Health and Clinical Medicine, Division of Medicine, Umeå University Hospital, Umeå, Sweden

Abstract

Our genes are the conduit through which the environment communicates to the cells in our bodies. The responses to these signals include hormonal, metabolic and neurological changes to tissues and organs that manifest as phenotypes – measurable responses to gene transcription and translation. Thus, the health consequences of lifestyle behaviors such as physical activity, which can be broadly defined as 'environmental' exposures, are channeled through our genes. The extent to which these signals are conveyed depends in part on the structure and function of our genome. Hence, even when exposed to the same exercise regimes or doses of physical activity, responses vary markedly from one person to the next; some experience marked changes in disease phenotypes such as lipid and glucose concentrations, adiposity, or blood pressure levels, whilst others appear unresponsive. It is this process that underlies the concept that we will discuss in this chapter, a concept termed gene-lifestyle interaction. The aims of this chapter are (a) to convey to the reader the fundamental principles of gene-lifestyle interaction; (b) to describe the historical basis to this area of research; (c) to explain how understanding gene-lifestyle interactions might enhance our knowledge of the molecular mechanisms of disease; (d) to speculate on the ways in which information of gene-lifestyle interactions might eventually facilitate disease prevention, and (e) to overview the published literature which has focused on obesity as an outcome.

Copyright © 2009 S. Karger AG, Basel

Obesity and type 2 diabetes are major global health concerns owing to their rapidly increasing prevalences and the severe burden they impose on affected individuals and healthcare systems. Both diseases share behavioral and genetic determinants, with obesity frequently preceding the onset of diabetes. Genome-wide association studies (GWASs) have spearheaded a major transition in the discovery of the genes that predispose to obesity and diabetes. For almost all common complex diseases, the number of confirmed genetic risk variants has increased manifold in the past 3 years. Prior to 2006, only a handful of polymorphisms had been equivocally associated with type

2 diabetes or common obesity [1]. Today, the identities of roughly 40 independent diabetes or obesity-predisposing loci are known, most of which have been discovered using GWASs. These discoveries have rejuvenated interest in the role of genetics in personalized medicine. However, the successful translation of this information into the clinical setting is predicated on two assumptions; the first is that the level of risk conveyed by a genetic variant is clinically meaningful and the second is that, if so, genetic risk can be offset through clinical intervention. To this end, the notion that prescription exercise regimens might one day be tailored to the genotype of the patient is becoming increasingly popular [2].

The purpose of this chapter is to introduce the reader to the concept of gene × environment interaction (also known as context-dependent genetic risk), to describe how genetic variants and lifestyle behaviors interact, and to provide a perspective on how this information might one day facilitate public health interventions to prevent or delay the onset of cardiometabolic diseases.

For much of the past 100,000 years, humans were hunter-gatherers, where the acquisition of food necessitated physical activity and where exposure to the extremes of heat or cold was common. These demands on energy metabolism imposed selective pressures that favored the survival and reproduction of lineages that were genetically enriched to thrive in physically demanding environments. By contrast to the vast majority of human evolution, the past century has witnessed a progressive decline in habitual physical activity levels and a relative overconsumption of foods rich in sugar and animal fats. Thus, for many of us, our lifestyles are massively disconnected from those which were common throughout evolution – those within which our genomes are designed to function. For many indigenous populations hailing from ancestral lines defined by migration, famine, and cold exposure, the genomic advantage gained throughout human evolution may predispose them to a much higher risk of developing metabolic and cardiovascular diseases when living modern industrialized lifestyles. This may help explain the high level of variation in the rates of obesity and type 2 diabetes across populations living comparable lifestyles, with the highest rates often occurring in aboriginal groups such as Pima Indians, Alaskan Natives, and Polynesian Islanders [3].

Because meaningful changes in the structure of the human genome occur over thousands of years, recent changes in lifestyle and not in genetics per se must have triggered the emergence of the complex diseases that are prevalent in many industrialized nations. This notion is supported by clinical trials, which show that with increased physical activity and improvements in dietary composition weight loss is achievable, cardiovascular risk profiles can be improved, and the rates at which type 2 diabetes occurs can be massively reduced [4, 5]; for most, adopting lifestyle behaviors similar to those of our ancestors reverses the progression of cardiometabolic disease. However, there is also considerable variability in the extent to which an individual responds to lifestyle intervention, with some being highly responsive and others maintaining similar risk profiles across the intervention period [6]. A fraction of this variability may be explained by gene-lifestyle interactions.

why SNPs associated with traits that interfere with reproduction are common is that historically they conveyed a survival advantage.

As described above, much of human evolution involved exposure to physically demanding environments where infection, thermic stress, periods of food deprivation, and a requirement to be physically active predominated. In this environmental context, an ability to lay down adipose reserves when food was plentiful would have been essential for survival. As illustrated by modern examples of lipodystrophy, anorexia nervosa, cachexia and athletic training, women with little or no adipose tissue are amenorrheic, infertile, and are prone to infection. Thus, given the important role adipose tissue plays in reproduction and survival, it is likely that the human genome is enriched with genes that promote fatness and inhibit leanness. Thus, obesity and its pathogenic consequences are natural responses to an environment that lacks physical stress.

How Might Information on Gene-Lifestyle Interactions Elucidate the Molecular Basis of Disease?

The effects of lifestyle exposures on disease risk are mediated by our genes. For example, specific nutrients (e.g. polyunsaturated fatty acids, sodium or zinc) and components of physical activity (e.g. shear stress, energy flux or hyperthermia) activate genes involved in cardiovascular or metabolic processes beneficial to health. By contrast, restriction of these exposures leads to genomic dysregulation. The identification of gene-lifestyle interactions sheds light on the genes involved in mediating and modifying the effects of specific lifestyle exposures (or treatments). Thus, bona fide examples of gene-lifestyle interactions may help identify specific drug, behavioral or nutritional targets for disease treatment or prevention.

How Might Information on Gene-Lifestyle Interactions Help Identify Those at High Risk of Disease?

It is becoming increasingly evident that the between-individual variability in the susceptibility to disease in a given environment is attributable in part to genetic variation. Figure 2 shows the interaction between the PPARG Pro12Ala genotype and dietary fat intake on change in visceral fat area during a clinical trial for diabetes prevention [16]; these data suggest that the effects of dietary fat composition on visceral fat accumulation are dependent on the genotype carried at the Pro12Ala locus. Those carrying the Ala12 allele in this study accumulated visceral adipose tissue at a greater rate than Pro12 homozygotes when the ratio of dietary polyunsaturated fats (PUFAs) was low; this contrasts the effect of the genotype on visceral fat accumulation when PUFA intake was high. This information might be used to identify people who are most

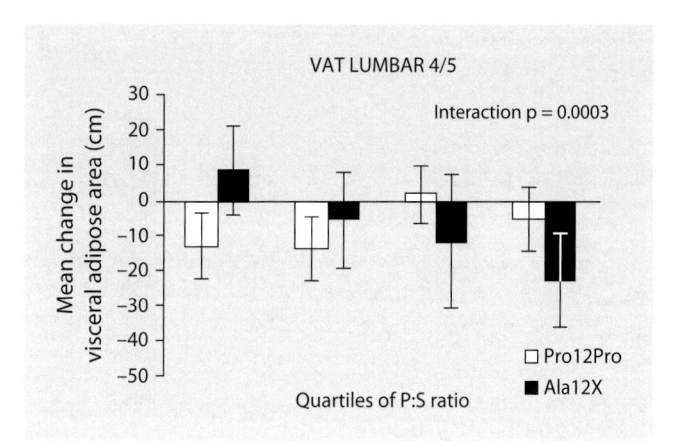

Fig. 2. Interaction between Pro12Ala genotypes and PUFA intake on 1-year change in visceral adiposity in the metformin treatment arm (n = 149). Data are means (95% CI). Adapted from Franks et al. [16].

susceptible to become obese or develop cardiovascular disorders when eating diets high in saturated fats. These data also identify a genetic subgroup of individuals who appear to benefit most in terms of visceral fat reduction when consuming a diet high in PUFAs, raising the possibility of genome-guided prevention through nutritional intervention.

How Might Information on Gene-Lifestyle Interactions Be Used in the Treatment of Disease?

We have studied the interaction between the E23K variant at the KCNJ11 (potassium inwardly-rectifying channel, subfamily J, member 11) gene and metformin treatment on diabetes risk reduction in the DPP [17]. The overall reduction in diabetes risk with metformin treatment at 3.2 years was 32% in the DPP. However, as shown in figure 3, the reduction in risk was confined largely to people carrying the E23E genotype. No beneficial effect of metformin was observed in those carrying one or both copies of the K23 allele in our study. These data suggest that in carriers of the K23 allele at KCNJ11, metformin is an ineffective therapy. One might use these data to guide the choice of glucose-lowering medications. Because metformin shares some features in common with exercise (e.g. activation of the AMPK pathway and weight loss in the absence of muscle atrophy), understanding interactions between gene variants and metformin might shed light on the mechanism through which exercise interactions with our genes.

Figure 4 shows the interaction between the G62A variant at the G-protein-coupled receptor 10 gene and physical activity on blood pressure levels [18]. In people carrying the A allele (approximately 40% of the population), no relationship between physical activity and blood pressure was apparent. By contrast, in those carrying the GG genotype, a strong inverse relationship was observed. These data suggest that the

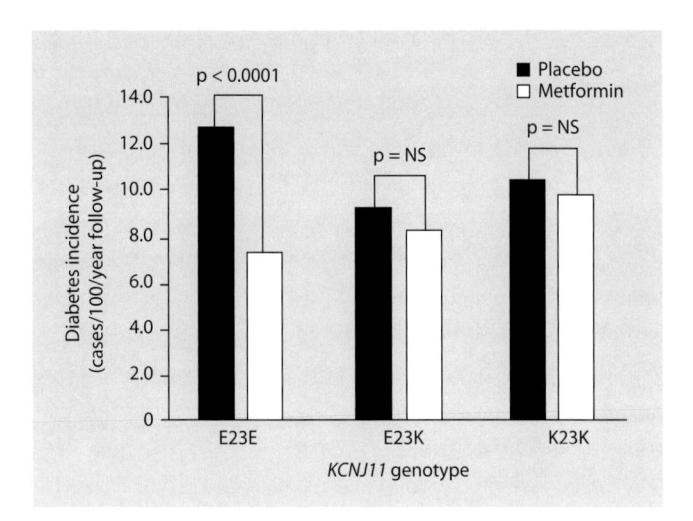

Fig. 3. Interaction between E23K genotypes and metformin treatment on diabetes incidence. Data are cumulative incidence. Adapted from Florez et al. [17].

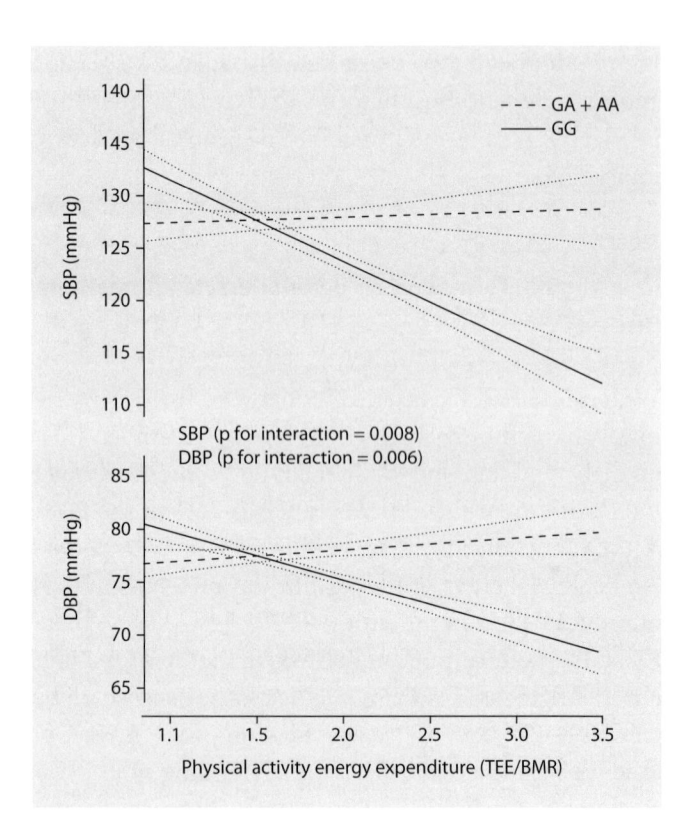

Fig. 4. Interaction between a polymorphism in the G-protein coupled receptor 10 gene and physical activity on blood pressure in 762 middle-aged adults from the UK. TEE/BMR = Total energy expenditure/basal metabolic rate. Reproduced from Franks et al. [18].

effects of physical activity on blood pressure may differ depending on genotype at this locus. Whilst physical activity has many other beneficial health effects and should rarely be discouraged, one might use this information on gene-lifestyle interaction to guide the choice of adjuvant antihypertensive therapies.

Overview of the Literature on Gene-Physical Activity Interactions on Human Obesity

The ability to undertake high-density genotyping in large population samples has facilitated a rapidly growing number of studies that report evidence of gene-exercise interactions on obesity-related traits (summarized in table 1). These studies include observational cohort, case-control, and intervention studies. Almost all studies published at the time of writing this chapter have focused on biological candidate genes or family-based linkage scans. Three studies in twins have also been reported [22–49]. Only a very small number of studies have tested variants identified through GWASs [22–51], which is probably owing to the recent emergence of these loci. These reports include a cohort study in which a variant at the FTO locus (rs9939609) was shown to interact with physical activity in around 5,000 middle-aged Danish adults [22]. In that study, the effect of the risk allele on obesity predisposition, which is now well established [15, 52], was clear in sedentary individuals, but was abolished in those who reported moderate to high levels of physical activity. In a subsequent cohort study set within around 700 members of the Amish community in Lancaster County, Pennsylvania, a second FTO gene variant (rs1861868), which is in weak linkage disequilibrium with the rs9939609 variant (r^2 ~0.17), also appeared to interact with physical activity (assessed using uniaxial accelerometry) [50]. Elsewhere, we examined whether the rs9939609 variant modifies the response to intensive lifestyle intervention in a randomized clinical trial of around 3,000 individuals at high risk of type 2 diabetes [53]. In that study, we did not find evidence to support an interaction between the rs9939609 variant and lifestyle intervention on changes in weight or BMI. However, we observed a modest interaction on 1-year change in visceral adipose volume determined through computed tomography. Although these studies might be considered confirmation of the initial Danish observation, it is important to stress that because the Amish study focused on a different FTO variant, and our study was set within the context of a clinical trial, the interaction was of borderline statistical significance and for a different obesity trait, appropriate confirmation studies are still required. To this end, we recently undertook a confirmation study focusing on the same variant and outcome studied in the Danish report [see 61]. Our study was undertaken in 16,000 Swedish adults living close to the region of Denmark in which the original report was set. The participants were of similar age and social status to the Danish participants. Nonetheless, we found no evidence to support the original study's finding that the rs9939609 variant and physical activity interact to influence obesity predisposition.

An interesting twist to the FTO story has been the recent discovery that persons carrying the rs9939609 risk alleles may be more physically active than those at lower genetic risk, despite being more obese. This observation was originally reported in Scottish children who underwent detailed assessments of energy intake and expenditure using doubly-labeled water [54], and confirmed in the Swedish cohort of 16,000 adults described above [55]. The explanation for this apparently paradoxical

Table 1. Overview of studies focusing on gene-physical activity interactions on obesity

Study	Study design	Description of cohort	PA assessment/ exposure
Observational studies			
Berentzen et al. [2004]	Prospective case-control study	• White males • Cases: n = 568; BMI 33.2 (2.3); age 32.0 (6.2) years • Controls: n = 717; BMI 21.5 (2.3); age 37.1 (8.6) years	Questionnaire
Franks PW et al. [2004]	Population-based cohort	Low physical activity level • Pro12Ala+ Ala/Ala Females: n = 26; BMI 25.5 (0.90); age 51.6 (1.79) years Males: n = 27; BMI 27.8 (0.64); age 54.6 (2.39) years • Pro12Pro Females: n = 114; BMI 26.9 (0.52); Age 53.9 (0.99) years Males: n = 86; BMI 27.0 (0.44); age 51.7 (1.26) years High physical activity level • Pro 12Ala+ Ala/Ala Females: n = 32; BMI 25.9 (0.68); age 48.8 (1.64) years Males: n = 22; BMI 26.5 (0.81); age 53.5 (2.63) years • Pro12Pro Females: n = 108; BMI 25.6 (0.42); age 51.7 (1.00) years Males: n = 91; BMI 26.9 (0.36); age 56.2 (1.01) years	4 days of free-living heart rate monitoring
Marti del Moral et al. [2002]	Case-control study	Cases: n = 159; BMI > 30 Controls: n = 154; BMI < 25; age 20–60 years	Questionnaire

ne (variant)	Outcomes	Significant	Main results
P2 (3′ UTR I/D) P3 (−55 C/T)	Change in BMI, waist, and fat mass	No	Neither PA nor the gene variants were consistently associated with changes in BMI; there was no evidence for statistical interactions between the UCP variants and physical activity in relation to changes in any of the obesity measures
ARG (Pro12Ala)	physical activity level, fasting insulin, polyunsaturated to saturated fatty acid ratio (P:S ratio)	Yes	In Pro allele homozygotes the combined associations of P:S ratio and PAL are additive; in contrast, in Ala allele carriers, PAL modified the association between P:S ratio and fasting insulin level in a multiplicative manner
RB3 (Trp64Arg)	BMI, M/S (metabolic equivalents divided with hours sitting down S) predict recreational energy expenditure	Yes	The risk of obesity associated with the ADRB3 Trp64Arg polymorphism may be dependent physical activity levels; the effect (defined as odds ratio) of the polymorphism on obesity was 2.98 was apparent only among subjects with levels of recreational energy expenditure above the population median

Table 1. Continued

Study	Study design	Description of cohort	PA assessment/exposure
Kilpeläinen et al. [2007]	Prospective cohort (4.1 years)	Three physical activity groups (n = 479) • Group 1: males/females, n = 53/n = 105; BMI 31.2 (4.1); age 56.1(7.0) years • Group 2: males/females, n = 50/n = 110; BMI 31.8 (5.1); age 53.9 (7.1) years • Group 3: males/ females, n = 56/n = 105; BMI 0.7(3.9); age 56.5(6.6) years	Questionnaire annually
Posion et al. [2001]	Cross-sectional study	African immigrants and African-American (n = 175) • CC or CT: n = 77 (88% females); BMI 29.9(6.5); age 42.4(7.5) years • TT: n = 98 (84.7% females); BMI 29.8 (4.9); age 42.8 (7.5) years	Self-reported physical activity (SRPA)
Moore et al. [2001]	Cross-sectional study	White postmenopausal non obese women (n = 63) • Sedentary: n = 19; BMI 23.7 (0.6); age 64 (1) years • Active: n = 20; BMI 25.2 (0.6); age 63 (1) years • Athletic: n = 24; BMI 21.7 (0.4); age 65 (1) years	VO_{2max}
Otabe et al. [2000]	Case-control study	• White obese females/males: n = 318/83; BMI 46.5 (7.5); age 44 (12) years • Control group females/males: n = 130/101; BMI 23(2.5); age 56 (14) years	Questionnaire
Ridderstråle et al. [2006]	Cross-sectional study	• Swedish men: n = 902; BMI 27.0 (4.6); age 48.0 (11.8) years • Swedish women: n = 899; BMI 26.8 (4.6); age 48.1 (11.9) years	Questionnaire

Gene (variant)	Outcomes	Significant	Main results
SLC2A2 (rs5393, rs5394, rs5404, rs5400) ABCC8 (rs2188966, rs1799859, rs3758947) KCNJ11(rs5219)	physical activity (METs), BMI, waist circumference, OGTT	Yes	Changes in moderate to vigorous physical activity during 4 years modified the effect of polymorphisms in SLC2A and ABCC8 on glucose levels and the conversion from IGT to type 2 diabetes among individuals with IGT; the authors concluded that the interaction of physical activity with genes regulating insulin secretion indirectly suggests that physical activity not only improves insulin sensitivity but also preserve β-cell function.
GNB3 (CC, CT, TT)	physical activity, BMI	No	No statistically significant effect of the genotypes or interaction between genotype and the level of physical activity on BMI.
ADRB2 (Gln27Glu)	body fat %, fat mass, VO_{2max}, weight, BMI	Yes	Study concluded that the major Gln27 allele may be associated with elite athletic performance
UCP3 (−55 C>T)	physical activity index, BMI	Yes	Study suggested that the UCP3 variant may modify the association between physical activity and obesity
PPARGC1A (Gly482Ser)	BMI	Yes	Elderly men ≥50 years carrying 482Ser had an increased risk of obesity compared with subjects who were homozygous for the major allele; the risk was restricted to males with a low leisure-time physical activity level

genotype, thus improving the effectiveness of the intervention. However, these prospects are contingent on the discovery, confirmation, and refinement of genetic loci that interact with lifestyle behaviors to influence disease risk and progression. Ideally, this evidence should include studies from appropriately conducted randomized controlled trials. Despite a burgeoning literature on gene-lifestyle interactions, there are few (if any) convincing examples of genetic loci that act in this way. Indeed, as we have previously argued, no appropriately designed intervention studies have been reported that test whether variability in phenotypic responses to exercise intervention is attributable to genetic variation [57]. Although studies such as HERITAGE clearly illustrate that the phenotypic response to exercise varies greatly from one person to the next and that this is related to inherited factors [6], nongenetic factors are also inherited [59]. Moreover, even though studies such as HERITAGE [6] and DREW [58] diligently administered the exercise intervention, physical activity (and other behaviors) outside the training sessions was not tightly controlled; these factors rather than genetic variation may explain why people responded differently to the same exercise regime [60]. Having said this, given what we understand of human evolution, it remains highly intuitive that genetic variation influences exercise response. Thus, to move from the stage of speculating about the combined role of genetics and exercise for health promotion and disease prevention to one where this becomes reality will require a paradigm shift. This means that the standards which have become commonplace elsewhere in human genetics research, where reports of genetic associations with disease traits undergo extensive replication efforts before being accepted as real, will need to be adopted by those of us focusing on gene-lifestyle interactions if our work is to meaningfully impact human health.

References

1 McCarthy MI, Abecasis GR, Cardon LR, Goldstein DB, Little J, Ioannidis JP, Hirschhorn JN: Genome-wide association studies for complex traits: consensus, uncertainty and challenges. Nat Rev Genet 2008;9:356–369.
2 Roth SM: Perspectives on the future use of genomics in exercise prescription. J Appl Physiol 2008;104:1243–1245.
3 King H, Rewers M: Global estimates for prevalence of diabetes mellitus and impaired glucose tolerance in adults. WHO Ad Hoc Diabetes Reporting Group. Diabetes Care 1993;16:157–177.
4 Knowler WC, Barrett-Connor E, Fowler SE, Hamman RF, Lachin JM, Walker EA, Nathan DM: Reduction in the incidence of type 2 diabetes with lifestyle intervention or metformin. N Engl J Med 2002;346:393–403.
5 Tuomilehto J, Lindstrom J, Eriksson JG, Valle TT, Hamalainen H, Ilanne-Parikka P, Keinanen-Kiukaanniemi S, Laakso M, Louheranta A, Rastas M, Salminen V, Aunola S, Cepaitis Z, Moltchanov V, Hakumaki M, Mannelin M, Martikkala V, Sundvall J, Uusitupa M, the Finnish Diabetes Prevention Study Group: Prevention of type 2 diabetes mellitus by changes in lifestyle among subjects with impaired glucose tolerance. N Engl J Med 2001;344:1343–1350.
6 Bouchard C, Rankinen T: Individual differences in response to regular physical activity. Med Sci Sports Exerc 2001;33:S446–S451.
7 Neel JV: Diabetes mellitus: a 'thrifty' genotype rendered detrimental by 'progress'? Am J Hum Genet 1962;14:353–362.

8 Chakravarthy MV, Booth FW: Eating, exercise, and 'thrifty' genotypes: connecting the dots toward an evolutionary understanding of modern chronic diseases. J Appl Physiol 2004;96:3–10.

9 Bersaglieri T, Sabeti PC, Patterson N, Vanderploeg T, Schaffner SF, Drake JA, Rhodes M, Reich DE, Hirschhorn JN: Genetic signatures of strong recent positive selection at the lactase gene. Am J Hum Genet 2004;74:1111–1120.

10 Voight BF, Kudaravalli S, Wen X, Pritchard JK: A map of recent positive selection in the human genome. PLoS Biol 2006;4:e72.

11 Speakman JR: A nonadaptive scenario explaining the genetic predisposition to obesity: the 'predation release' hypothesis. Cell Metabolism 2007;6:5–12.

12 Speakman JR: Thrifty genes for obesity and the metabolic syndrome – time to call off the search? Diab Vasc Dis Res 2006;3:7–11.

13 Speakman JR: Thrifty genes for obesity, an attractive but flawed idea, and an alternative perspective: the 'drifty gene' hypothesis. Int J Obes 2008;32:1611–1617; http://dx.doi.org/10.1038/ijo.2008.161.

14 Prentice A, Hennig BJ, Fulford AJ: Evolutionary origins of the obesity epidemic: natural selection of thrifty genes or genetic drift following predation release? Int J Obes 2008;32:1607–1610.

15 Frayling TM, Timpson NJ, Weedon MN, et al: A common variant in the FTO gene is associated with body mass index and predisposes to childhood and adult obesity. Science 2007;316:889–894.

16 Franks PW, Jablonski KA, Delahanty L, Hanson RL, Kahn SE, Altshuler D, Knowler WC, Florez JC: The Pro12Ala variant at the peroxisome proliferator-activated receptor gamma gene and change in obesity-related traits in the Diabetes Prevention Program. Diabetologia 2007;50:2451–2460.

17 Florez JC, Jablonski KA, Kahn SE, Franks PW, Dabelea D, Hamman RF, Knowler WC, Nathan DM, Altshuler D: Type 2 diabetes-associated missense polymorphisms KCNJ11 E23K and ABCC8 A1369S influence progression to diabetes and response to interventions in the Diabetes Prevention Program. Diabetes 2007;56:531–536.

18 Franks PW, Bhattacharyya S, Luan J, Montague C, Brennand J, Challis B, Brage S, Ekelund U, Middelberg RP, O'Rahilly S, Wareham NJ: Association between physical activity and blood pressure is modified by variants in the G-protein coupled receptor 10. Hypertension 2004;43:224–228.

19 Ueno LM, Frazzatto ES, Batalha LT, Trombetta IC, do Socorro Brasileiro M, Irigoyen C, Brum PC, Villares SM, Negrão CE: Alpha2B-adrenergic receptor deletion polymorphism and cardiac autonomic nervous system responses to exercise in obese women. Int J Obes (Lond) 2006;30:214–220.

20 Thamer C, Machann J, Stefan N, Schäfer SA, Machicao F, Staiger H, Laakso M, Böttcher M, Claussen C, Schick F, Fritsche A, Haring HU: Variations in PPARD determine the change in body composition during lifestyle intervention: a whole-body magnetic resonance study. J Clin Endocrinol Metab 2008;93:1497–1500.

21 Naito H, Kamijima M, Yamanoshita O, Nakahara A, Katoh T, Tanaka N, Aoyama T, Gonzalez FJ, Nakajima T: Differential effects of aging, drinking and exercise on serum cholesterol levels dependent on the PPARA-V227A polymorphism. J Occup Health 2007;49:353–362.

22 Andreasen CH, Stender-Petersen KL, Mogensen MS, Torekov SS, Wegner L, Andersen G, Nielsen AL, Albrechtsen A, Borch-Johnsen K, Rasmussen SS, Clausen JO, Sandbaek A, Lauritzen T, Hansen L, Jorgensen T, Pedersen O, Hansen T: Low physical activity accentuates the effect of the FTO rs9939609 polymorphism on body fat accumulation. Diabetes 2008;57:95–101.

23 Rosado EL, Bressan J, Martins MF, Cecon PR, Martínez JA: Polymorphism in the PPARgamma2 and beta2-adrenergic genes and diet lipid effects on body composition, energy expenditure and eating behavior of obese women. Appetite 2007;49:635–643.

24 Kilpeläinen TO, Lakka TA, Laaksonen DE, Laukkanen O, Lindström J, Eriksson JG, Valle TT, Hämäläinen H, Aunola S, Ilanne-Parikka P, Keinänen-Kiukaanniemi S, Tuomilehto J, Uusitupa M, Laakso M, Finnish Diabetes Prevention Study Group: Physical activity modifies the effect of SNPs in the SLC2A2 (GLUT2) and ABCC8 (SUR1) genes on the risk of developing type 2 diabetes. Physiol Genomics 2007;31:264–272.

25 Wootton PT, Flavell DM, Montgomery HE, World M, Humphries SE, Talmud PJ: Lipoprotein-associated phospholipase A2 A379V variant is associated with body composition changes in response to exercise training. Nutr Metab Cardiovasc Dis 2007;17:24–31.

26 Okura T, Rankinen T, Gagnon J, Lussier-Cacan S, Davignon J, Leon AS, Rao DC, Skinner JS, Wilmore JH, Bouchard C: Effect of regular exercise on homocysteine concentrations: the HERITAGE Family Study. Eur J Appl Physiol 2006;98:394–401.

27 Karnehed N, Tynelius P, Heitmann BL, Rasmussen F: Physical activity, diet and gene-environment interactions in relation to body mass index and waist circumference: the Swedish young male twins study. Public Health Nutr 2006;9:851–858.

28 de Luis Roman D, de la Fuente RA, Sagrado MG, Izaola O, Vicente RC: Leptin receptor Lys656Asn polymorphism is associated with decreased leptin response and weight loss secondary to a lifestyle modification in obese patients. Arch Med Res 2006;37:854–859.

29 Ridderstråle M, Johansson LE, Rastam L, Lindblad U: Increased risk of obesity associated with the variant allele of the PPARGC1A Gly482Ser polymorphism in physically inactive elderly men. Diabetologia 2006;49:496–500.

30 Thompson PD, Tsongalis GJ, Ordovas JM, Seip RL, Bilbie C, Miles M, Zoeller R, Visich P, Gordon P, Angelopoulos TJ, Pescatello L, Moyna N: Angiotensin-converting enzyme genotype and adherence to aerobic exercise training. Prev Cardiol 2006;9:21–24.

31 Berentzen T, Dalgaard LT, Petersen L, Pedersen O, Sørensen TI: Interactions between physical activity and variants of the genes encoding uncoupling proteins-2 and -3 in relation to body weight changes during a 10-y follow-up. Int J Obes (Lond) 2005;29:93–99.

32 Tworoger SS, Chubak J, Aiello EJ, Yasui Y, Ulrich CM, Farin FM, Stapleton PL, Irwin ML, Potter JD, Schwartz RS, McTiernan A: The effect of CYP19 and COMT polymorphisms on exercise-induced fat loss in postmenopausal women. Obes Res 2004;12:972–981.

33 Franks PW, Luan J, Browne PO, Harding AH, O'Rahilly S, Chatterjee VK, Wareham NJ: Does peroxisome proliferator-activated receptor gamma genotype (Pro12ala) modify the association of physical activity and dietary fat with fasting insulin level? Metabolism 2004;53:11–16.

34 Walston J, Andersen RE, Seibert M, Hilfiker H, Beamer B, Blumenthal J, Poehlman ET: Arg64 beta3-adrenoceptor variant and the components of energy expenditure. Obes Res 2003;11:509–511.

35 Hittel DS, Kraus WE, Hoffman EP: Skeletal muscle dictates the fibrinolytic state after exercise training in overweight men with characteristics of metabolic syndrome. J Physiol 2003;548:401–410.

36 Marti A, Corbalán MS, Martínez-Gonzalez MA, Martinez JA: TRP64ARG polymorphism of the beta 3-adrenergic receptor gene and obesity risk: effect modification by a sedentary lifestyle. Diabetes Obes Metab 2002;4:428–430.

37 Kahara T, Takamura T, Hayakawa T, Nagai Y, Yamaguchi H, Katsuki T, Katsuki K, Katsuki M, Kobayashi K: Prediction of exercise-mediated changes in metabolic markers by gene polymorphism. Diabetes Res Clin Pract 2002;57:105–110.

38 Poston WS, Haddock CK, Spertus J, Catanese DM, Pavlik VN, Hyman DJ, Hanis CL, Forevt JP: Physical activity does not mitigate G-protein-related genetic risk for obesity in individuals of African descent. Eat Weight Disord 2002;7:68–71.

39 Moore GE, Shuldiner AR, Zmuda JM, Ferrell RE, McCole SD, Hagberg JM: Obesity gene variant and elite endurance performance. Metabolism 2001;50:1391–1392.

40 Chagnon YC, Rice T, Pérusse L, Borecki IB, Ho-Kim MA, Lacaille M, Paré C, Bouchard L, Gagnon J, Leon AS, Skinner JS, Wilmore JH, Rao DC, Bouchard C: HERITAGE Family Study. Genomic scan for genes affecting body composition before and after training in Caucasians from HERITAGE. J Appl Physiol 2001;90:1777–1787.

41 Otabe S, Clement K, Dina C, Pelloux V, Guy-Grand B, Froguel P, Vasseur F: A genetic variation in the 5′ flanking region of the UCP3 gene is associated with body mass index in humans in interaction with physical activity. Diabetologia 2000;43:245–249.

42 Feitosa MF, Rice T, Nirmala A, Rao DC: Major gene effect on body mass index: the role of energy intake and energy expenditure. Hum Biol 2000;72:781–799.

43 Rice T, Hong Y, Pérusse L, Després JP, Gagnon J, Leon AS, Skinner JS, Wilmore JH, Bouchard C, Rao DC: Total body fat and abdominal visceral fat response to exercise training in the HERITAGE Family Study: evidence for major locus but no multifactorial effects. Metabolism 1999;48:1278–1286.

44 Boer JM, Kuivenhoven JA, Feskens EJ, Schouten EG, Havekes LM, Seidell JC, Kastelein JJ, Kromhout D: Physical activity modulates the effect of a lipoprotein lipase mutation (D9N) on plasma lipids and lipoproteins. Clin Genet 1999;56:158–163.

45 Samaras K, Kelly PJ, Chiano MN, Spector TD, Campbell LV: Genetic and environmental influences on total-body and central abdominal fat: the effect of physical activity in female twins. Ann Intern Med 1999;130:873–882.

46 Meirhaeghe A, Helbecque N, Cottel D, Amouyel P: Beta2-adrenoceptor gene polymorphism, body weight, and physical activity. Lancet 1999;353:896.

47 Bouchard C, Tremblay A: Genetic effects in human energy expenditure components. Int J Obes 1990;14(suppl 1):49–55; discussion 55–58.

48 Yoshida T, Sakane N, Umekawa T, Sakai M, Takahashi T, Kondo M: Mutation of beta 3-adrenergic-receptor gene and response to treatment of obesity. Lancet 1995;346:1433–1434.

49 Ortega-Alonso A, Sipilä S, Kujala UM, Kaprio J, Rantanen T: Body fat and mobility are explained by common genetic and environmental influences in older women. Obesity (Silver Spring) 2008;16:1616–1621.

50 Rampersaud E, Mitchell BD, Pollin TI, Fu M, Shen H, O'Connell JR, Ducharme JL, Hines S, Sack P, Naglieri R, Shuldiner AR, Snitker S: Physical activity and the association of common FTO gene variants with body mass index and obesity. Arch Intern Med 2008;168:1791–1797.

51 Franks PW, Rolandsson O, Debenham SL, Fawcett KA, Payne F, Dina C, Froguel P, Mohlke KL, Willer C, Olsson T, Wareham NJ, Hallmans G, Barroso I, Sandhu MS: Replication of the association between variants in WFS1 and risk of type 2 diabetes in European populations. Diabetologia 2008;51:458–463.

52 Dina C, Meyre D, Gallina S, Durand E, Korner A, Jacobson P, Carlsson LM, Kiess W, Vatin V, Lecoeur C, Delplanque J, Vaillant E, Pattou F, Ruiz J, Weill J, Levy-Marchal C, Horber F, Potoczna N, Hercberg S, Le Stunff C, Bougneres P, Kovacs P, Marre M, Balkau B, Cauchi S, Chevre JC, Froguel P: Variation in FTO contributes to childhood obesity and severe adult obesity. Nat Genet 2007;39:724–726.

53 Franks PW, Jablonski KA, Delahanty LM, McAteer JB, Kahn SE, Knowler WC, Florez JC: Assessing gene-treatment interactions at the FTO and INSIG2 loci on obesity-related traits in the Diabetes Prevention Program. Diabetologia 2008;51:2214–2223.

54 Cecil JE, Tavendale R, Watt P, Hetherington MM, Palmer CNA: An obesity-associated FTO gene variant and increased energy intake in children. N Engl J Med 2008;359:2558–2566.

55 Jonsson A, Franks PW: Obesity, FTO gene variant, and energy intake in children. N Engl J Med 2009;360:1571–1572.

56 Wong MY, Day NE, Luan JA, Chan KP, Wareham NJ: The detection of gene-environment interaction for continuous traits: should we deal with measurement error by bigger studies or better measurement? Int J Epidemiol 2003;32:51–57.

57 Brito EC, Franks PW: Commentary on viewpoint: perspective on the future use of genomics in exercise prescription. J Appl Physiol 2008;104:1248.

58 Church TS, Earnest CP, Skinner JS, Blair SN: Effects of different doses of physical activity on cardiorespiratory fitness among sedentary, overweight or obese postmenopausal women with elevated blood pressure: a randomized controlled trial. JAMA 2007;297:2081–2091.

59 Franks PW: Muscling in on the genetics of quantitative disease traits. J Appl Physiol 2007;103:1111–1112.

60 Goran MI, Poehlman ET: Endurance training does not enhance total energy expenditure in healthy elderly persons. Am J Physiol 1992;263:E950–E957.

61 Jonsson A, Renström F, Lyssenko V, Brito EC, Isomaa B, Berglund G, Nilsson PM, Groop L, Franks PW: Assessing the interaction between FTO variation (rs9939609) and physical activity on obesity in 15,931 Swedish and 2,511 Finnish adults. Diabetologia 2009;52:1334–1338.

Dr. Paul W. Franks
Genetic Epidemiology and Clinical Research Group, Department of Public Health and Clinical Medicine
Division of Medicine, Umeå University Hospital
SE–90187 Umeå (Sweden)
Tel. +46 90 785 3354, Fax +46 90 137 633, E-Mail paul.franks@medicin.umu.se

Genetic Risk Factors for Musculoskeletal Soft Tissue Injuries

Malcolm Collins[a,b] · Stuart M. Raleigh[c]

MRC/UCT Research Unit for Exercise Science and Sports Medicine of [a]the South African Medical Research Council and [b]the Department of Human Biology, Faculty of Health Sciences, University of Cape Town, Cape Town, South Africa; [c]Division of Health and Life Sciences, University of Northampton, Northampton, UK

Abstract

Acute and overuse musculoskeletal soft tissues injuries are common as a result of participating in specific physical or workplace activities. Multiple risk factors, including genetic factors, are implicated in the aetiology of these injuries. Common musculoskeletal soft tissue injuries for which a genetic contribution has been proposed include the Achilles tendon in the heel, the rotator cuff tendons in the shoulder and the cruciate ligaments in the knee. Recent developments in the identification of genetic risk factors for tendon and ligament injuries will be reviewed. Sequence variants within genes that encode for several tendon and/or ligament extracellular matrix proteins have been shown to be associated with specific musculoskeletal soft tissues injuries. Variants within the *TNC*, *COL5A1* and *MMP3* genes co-segregate with chronic Achilles tendinopathy. The variant within the *TNC* gene also appears to co-segregate with Achilles tendon ruptures, while sequence variants within the *COL1A1* and *COL5A1* genes have been shown to be associated with cruciate ligament ruptures and/or shoulder dislocations. Whether these variants are directly involved in the development of these musculoskeletal soft tissue abnormalities or in strong linkage disequilibrium with actual disease-causing loci remains to be established. We proposed that genetic risk factors will in the future be included in multifactorial models developed to understand the molecular mechanisms that cause musculoskeletal soft tissue injuries or related pathology. Clinicians could eventually use these models to develop personalised training programmes to reduce the risk of injury as well as to develop treatment and rehabilitation regimens for the injured individual.

Copyright © 2009 S. Karger AG, Basel

Tendons and ligaments are common structures that can be injured as a result of participating in physical activity or specific activities in the workplace [1]. Collectively, these injuries, together with injuries to skeletal muscle, are referred to as musculoskeletal soft tissue injuries and can be broadly classified as either acute or overuse injuries [2]. Acute injuries result from a specific macrotraumatic event, while overuse injuries result from cumulative microscope damage to tissues caused over time by repetitive microtraumatic events. Although initially asymptomatic, the increased volume of tissue damage eventually becomes symptomatic [2].

Common musculoskeletal soft tissue injuries which will be covered in this review include the Achilles tendon in the heel, the rotator cuff tendons in the shoulder and the cruciate ligaments in the knee. Rupture of the Achilles tendon and anterior cruciate ligament (ACL) are examples of acute injuries, while Achilles tendinopathy (also historically referred to as Achilles tendonitis) is an example of an overuse injury. As much as 50% of all sporting injuries involve tendons of which up to a fifth can comprise Achilles tendon injuries, while the lifetime prevalence of Achilles tendon injuries has been reported to be as high as 11% [3–5]. Although the incidence of ACL injuries is low in the general population [6, 7], regular participation in sports, particularly those which involve a sudden deceleration or change in direction, may place an individual at up to ten times greater risk of ACL rupture [8]. ACL ruptures are considered one of the most severe injuries sustained in a sporting population [9].

Although the genetic basis of tendon and ligament injuries has been reviewed [10, 11], this chapter will focus on the more recent developments in the field.

Tendon and Ligament Structure

Tendons and ligaments are collagenous structures which have a similar composition and hierarchical structure [12–14]. Approximately two thirds of these tissues consist of water. The rest is made up of tightly packed parallel bundles consisting predominately of type I collagen fibrils (60–80% of the dry mass of tendons). The fibrils aggregate into larger units of fibres and fascicles which finally produce the tendon or ligament [14].

Other collagen types, such as types III–VI, XII and XIV, are also structural components of the fibrils and surrounding extracellular matrix (ECM; fig. 1) [13, 15–18]. Various proteoglycans, such as decorin, lumican and versican, and glycoproteins, such as elastin, tenascin C and COMP, are also important components of tendons and/or ligaments [13–15, 19, 20]. The quantity of these other collagens, proteoglycans and glycoproteins within these tissues are however significantly less than type I collagen. Regions within tendons which have to also resist, in addition to the usual tensile forces, compressive and shear forces, also contain fibrocartilaginous-specific proteins such as types II, and IX–XI collagen, as well as the proteoglycans, biglycan and aggrecan [17, 18]. The genes that encode for these proteins are located throughout the human genome [10]. Changes in the collagen, proteoglycan and/or glycoprotein content and composition, as well the expression of many of the genes that encode for these proteins have been shown to be altered in tendinopathy [16, 18, 21, 22] and ligament injuries [23].

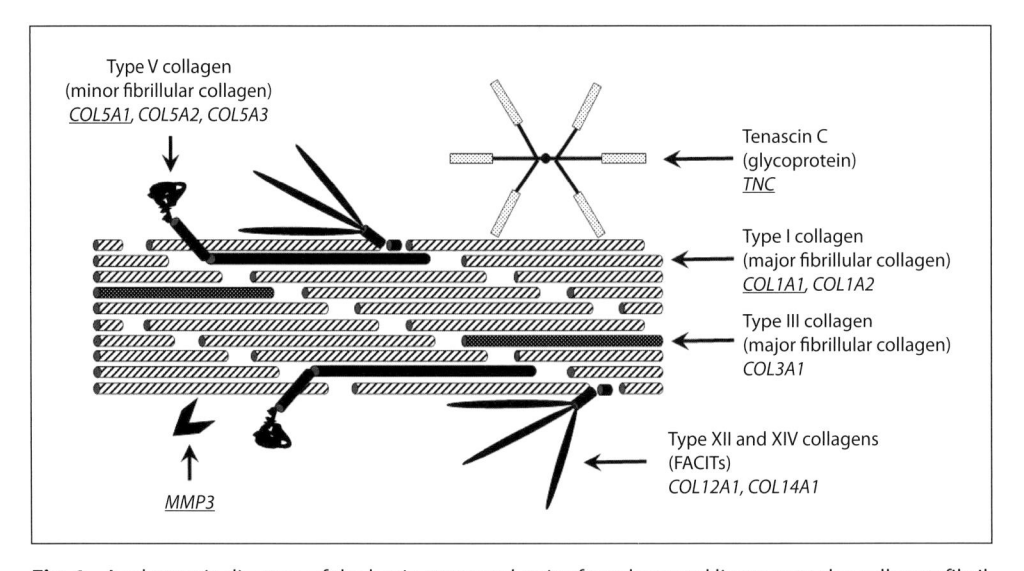

Fig. 1. A schematic diagram of the basic structural unit of tendons and ligaments, the collagen fibril. The fibril consists predominately of type I collagen. Type I collagen is a heterotrimer consisting of two α1(I) and one α2(I) chains, which are encoded for by the *COL1A1* (human chromosomal location 17q) and *COL1A2* (7q) genes, respectively. The fibril also contains trace amounts of types III and V collagen. Type III collagen is a homotrimer consisting of three α1(III) chains encoded for by the *COL3A1* (2q) gene. The major isoform of type V collagen is a heterotrimer consisting of two α1(V) and one α2(V) chains, which are encoded for by the *COL5A1* (9q) and *COL5A2* (2q) genes, respectively. Some isoforms contain an α3(V) chain encoded for by the *COL5A3* gene (19q). Types I and III collagen are members of the major fibrillar collagen sub-family, while type V collagen is a member of the minor fibrillar collagen sub-family. Types XII and XIV collagen are associated with the surface of the fibril and belong to the sub-family of fibril-associated collagens with interrupted triple helices (FACITs). Both proteins are homotrimers which are encoded for by the *COL12A1* (6q) and *COL14A1* (8q) genes. The hexameric glycoprotein, tenascin C and MMP3 are also shown. Tenascin C and MMP3 are encoded for by the *TNC* (9q) and *MMP3* (11q) genes, respectively. Other collagen types, proteoglycans and glycoproteins which are associated with or structural components of the fibril are not shown. The genes that have been shown to be associated with any musculoskeletal soft tissue injuries are underlined. The proteins are not necessarily drawn to scale.

Mechanism of Injury

The exact mechanisms responsible for tendon and ligament injuries are poorly understood [17, 24]. Since multiple risk factors are implicated in the aetiology of these injuries, these injuries are considered multifactorial disorders [17, 24]. The risk factors are divided into those which influence the athlete from without (extrinsic) and those from within (intrinsic) [2]. It has been proposed that the intrinsic risk factors predispose the athletes to a specific injury. Once predisposed, susceptibility to injury is determined by exposure to extrinsic risk factors. It is important to note that these intrinsic and extrinsic risk factors do not cause or necessarily result in an injury. The occurrence of a specific inciting event eventually results in an injury (fig. 2). The

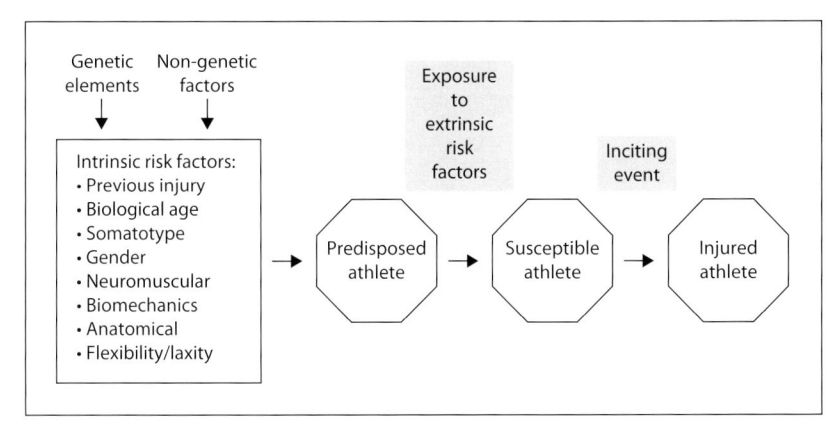

Fig. 2. A schematic diagram illustrating the complex relationship between intrinsic and extrinsic risk factors, as well as the role of the inciting event in the aetiology of soft tissue injuries [2, 25, 26]. Many intrinsic risk factors are, together with several non-genetic factors, determined to a lesser or greater extent, by genetic elements and individually could also be considered as multifactorial phenotypes.

relationship between intrinsic and extrinsic risk factors, as well as the role of the inciting event in the aetiology of sports injuries has been extensively reviewed [2, 25, 26].

Genetic factors have been suggested as intrinsic risk factors for Achilles tendon [27–30], rotator cuff tendon [31], shoulder dislocations [32] and ACL [32–36] acute and/or chronic injuries. Although the sequences of the 3.2 billion bases of human DNA are over 99.9% identical, the sequence variations (polymorphisms) in human DNA contribute to the visible and measurable biological variation observed in individuals. In fact, many of the common intrinsic risk factors implicated in injuries such as previous injury, somatotype, neuromuscular characteristics, biomechanical features, anatomical features and flexibility/laxity are, together with several other non-genetic factors, to a lesser or greater extent, determined by genetic factors [37]. Many of the intrinsic risk factors could therefore also in their own right be considered as multifactorial phenotypes. We hypothesize that polymorphisms within specific genes are partially responsible for the observed inter-individual variation in susceptibility to musculoskeletal soft tissue injuries.

Type I Collagen Genes and Musculoskeletal Soft Tissue Injuries

As previously mentioned, type I collagen is the major protein component of tendons and ligaments [14]. The protein is a heterotrimer consisting of two α1(I) and one α2(I) chains, which are encoded for by the *COL1A1* and *COL1A2* genes, respectively (fig. 1) [38]. Mutations within the *COL1A1* gene have been shown to cause connective tissue disorders such as osteogenesis imperfecta, and Ehlers-Danlos syndrome

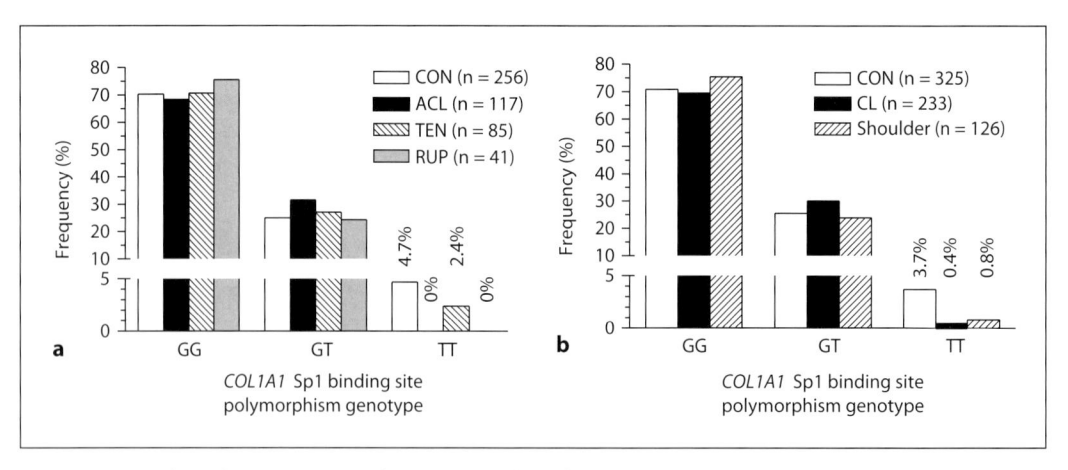

Fig. 3. The relative genotype frequencies of the functional Sp1 binding site polymorphism within intron 1 of the *COL1A1* gene in the South African control (CON), ACL rupture, chronic Achilles tendinopathy (TEN) and Achilles rupture (RUP) groups (**a**), as well as the Swedish CON and cruciate ligament (CL) rupture and shoulder dislocation (Shoulder) groups (**b**). The data from both South African Caucasian control populations, which consisted of unrelated individuals, were combined and presented [35, 46]. **a** CON versus ACL, OR = 12.0, 95% CI 0.7–204.8, p = 0.022; CON versus TEN, OR = 2.0, 95% CI 0.4–9.3, p = 0.531; CON versus RUP, OR = 4.2, 95% CI 0.2–73.1, p = 0.383. **b** CON versus CL, OR = 8.9, 95% CI 1.1–68.9, p = 0.010; CON versus Shoulder, OR = 4.8, 95% CI 0.6–37.3, p = 0.123.

[39]. Various single nucleotide polymorphisms within *COL1A1* have also been shown to be associated with several complex connective tissue disorders [40–44]. The G to T substitution in an intronic Sp1 binding site (rs1800012), resulting in increased affinity for the transcription factor Sp1, and increased gene expression, has been one of the most extensively investigated polymorphism within *COL1A1* [45]. With respect to musculoskeletal soft tissue injuries, the association of this polymorphism with cruciate ligament ruptures [32, 35], shoulder dislocation [32], Achilles tendon ruptures [46] and Achilles tendinopathy [46] has been investigated (fig. 3).

The rare TT genotype was found to be absent in South African Caucasian subjects with ACL (n = 117) [35] or Achilles tendon (n = 41) [46] ruptures (fig. 3a). In a Swedish study, Khoschnau et al. [32] only identified one subject each with a TT genotype with cruciate ligament ruptures (0.4%, n = 233) or shoulder dislocations (0.8%, n = 126; fig. 3b). Two subjects with the rare TT genotype were however identified in the subjects with chronic Achilles tendinopathy (2.4%, n = 85) [46] (fig. 3a). Similar TT genotype frequencies were reported in the control subjects in both the South African (4.7%, n = 256) and Swedish (3.7%, n = 325) studies, which is in agreement with the genotype frequencies reported in larger control populations [45, 47]. When combined, the rare TT genotype was significantly underrepresented in the subjects with either ligament or tendon ruptures (0.4%, n = 2 of 517) when compared with the control subjects (4.1%, n = 24 of 581, p < 0.0001) [48]. Although the rare TT genotype appears to have a protective role against acute soft tissue ruptures, larger sample sizes should be evaluated in the future to further investigate the effect of this genotype on various ligament, tendon and other soft tissue ruptures.

It is also interesting to note that, in the South African study, subjects with an ACL rupture were more than four times as likely to have a blood relative with a history of a ligament injury when compared with the control subjects [35]. This supports previous research that reported a congenital component [33] and familial genetic predisposition [34] to ACL injuries. It is however possible that the family members of the ACL subjects in the South African study have a higher exposure to non-genetic risk factors, such as participation in sports which involve a sudden change in direction or deceleration, compared with the family members of the control subjects.

The possible association of any other polymorphisms within the 15.8 kb *COL1A1* gene, which consists of 51 exons, or any polymorphisms within the 52 exons and 51 introns of the 36.3 kb *COL1A2* gene with soft tissue injuries has to date not been investigated.

Soft Tissue Injuries and Chromosome 9

It is highly unlikely that a single gene or a single variant within one gene exclusively predisposes athletes to musculoskeletal soft tissue injuries. These injuries, like other multifactorial disorders and phenotypes [11, 49, 50], are more likely polygenic in nature. However, which of the other 20,000–25,000 protein-coding genes located throughout the human genome are associated with these injuries? Although beyond the scope of this review, it is also important to mention that non-protein-coding genes could also be considered as possible genetic risk factors for these injuries.

Technologies have been developed and are available to screen the whole human genome in a single experiment for identifying the genetic differences between cohorts [51]. Although one advantage of this technology is obtaining large amounts of genetic information, it requires large sample sizes, is expensive and is often not hypothesis driven. Another approach for identifying genes associated with musculoskeletal soft tissue injuries is to select and test specific candidate genes based on their known biological function [10, 11]. Although this approach also has certain limitations [52, 53], it requires smaller sample sizes, is hypothesis driven and, as reviewed by September et al. [10, 11], it has successfully been used to identify and test candidate genes for Achilles tendon injuries based on their biological function and chromosomal location.

Some investigators have reported an association of the ABO blood group with Achilles tendon injury [10]. Furthermore, the ABO blood group is determined by a single gene located on the end of the long arm of chromosome 9 (9q34). Therefore, based on their biological function and chromosomal location the *COL5A1* and *TNC* genes were selected as candidate genes for Achilles tendon injuries [29, 30]. The *COL5A1* and *TNC* genes encode for the α1 chain of type V collagen and tenascin C, respectively [10, 11]. Type V collagen molecules intercalate into the tendon and ligament fibrils where they are believed, together with other proteins, to play an important role in regulating fibrillogenesis and modulating fibril growth [15, 17, 54] (fig. 1). Tenascin-C, on the other hand, is localised predominantly in regions responsible for transmitting high levels of mechanical force within tendons and ligaments [55]. The expression of the gene is regulated by mechanical loading [55–57].

It was found that a GT dinucleotide repeat polymorphism within intron 17 of the *TNC* gene is associated with both chronic Achilles tendinopathy and Achilles tendon rupture. Variants containing 12 and 14 GT repeats were overrepresented in individuals with either injury, while variants containing 13 and 17 repeats were underrepresented. In addition to the *TNC* gene, a *Bst*UI restriction fragment length polymorphism (RFLP) (rs12722) within the 3′-untranslated region of the *COL5A1* gene was also associated with chronic Achilles tendinopathy in South African [30] and Australian populations [58]. In both populations, the CC genotype was underrepresented by greater than 2-fold in the tendinopathy patients. Although only investigated in a small cohort, the *COL5A1 Bst*UI RFLP does not appear to be associated with Achilles tendon ruptures [30]. More recently, Posthumus et al. [36] have shown that the CC genotype of this polymorphism was also significantly underrepresented in female, but not male, athletes with ACL ruptures. The possible role(s) of type V collagen and/or tenascin-C in the development of soft tissue injuries needs to be investigated.

Other Matrix Genes

Since associations of the *COL5A1* and *TNC* genes with Achilles tendon and/or ACL injuries have already been described, any gene that encodes for other ECM proteins involved in similar biological processes as type V collagen and tenascin-C within these tissues would also be an ideal candidate gene for musculoskeletal soft tissue injuries. For example, proteins, such as decorin [59], lumican [60], fibromodulin [60], thrombospondin 2 [61] and types XII and XIV collagen [62–64] regulate fibrillogenesis and, therefore, their genes could also be investigated as genetic markers for these injuries.

Besides tenascin-C, the homotrimeric types XII and XIV collagen, which are also expressed in both tendons and ligaments, are regulated by mechanical loading [55, 65] (fig. 1). In particular, the small XIIB-1 isoform of type XII collagen is predominately expressed in tendons and ligaments [66]. These molecules belong to the subfamily of fibril-associated collagens with interrupted triple helices (FACIT) [67–69] which are believed to form interfibrillar connections and mediate fibril interaction with other extracellular and cell surface molecules [70–73] (fig. 1). Sequence variants within the *COL12A1* and *COL14A1* genes, which encode for types XII and XIV collagen, respectively, were individually not associated with Achilles tendinopathy [74].

Although not significant, the absence of both the rare CC and GG genotypes of the *COL12A1 Bsr*I and *Alu*I RFLPs, respectively, in the Achilles tendon rupture subjects warrants further research [74]. The AA genotype of the *COL12A1 Alu*I RFLP has recently been shown to be significantly overrepresented in female, but not male, subjects with ACL ruptures [see 99].

Matrix Metalloproteinases and ADAMTSs

The integrated nature of biological systems implies that multiple genes which encode for other structural components of tendons and ligaments or biological processes, such as adaptation, healing and remodelling, are probably also associated with these injuries. Besides the genes which encode for the ECM proteins already discussed, the genes that encode for any protein that regulates tendon and ligament biology,such as matrix metalloproteinases (MMPs), ADAMTSs (a disintegrin like and metalloprotease-thrombospondin with type 1 motif), tissue inhibitor of metalloproteinases, growth factors, cytokines and interleukins involved in musculoskeletal soft tissue biology are also ideal candidates [11]. Although all these families of proteins are important, only the MMPs and ADAMPTs will be further discussed in this review.

The MMPs are a family of at least 20 distinct endopeptidases that catalyse a plethora of ECM- and non-ECM-associated substrates [75]. MMPs have important roles in a variety of biological processes including wound healing [76], angiogenesis [77] and the regulation of various types of tissue turnover [78]. Although the genes that encode for the MMPs have been mapped throughout the human genome, many form a cluster on chromosome 11 (www.ncbi.nlm.nih.gov).

There is ample evidence to show that the expression of several *MMP*s is altered in damaged tendons [79, 80]. Furthermore, it has recently been demonstrated that expression of *MMP3* and *MMP13* is influenced by changes in mechanical loading in the Achilles and supraspinatus tendons [81]. In addition, serum levels of both MMP2 and MMP7 appear to be raised in individuals with a history of Achilles tendon rupture [82], and synovial fluid collected from the knees of individuals with ACL injuries is reported to have elevated levels of the MMP3 protein [83].

Alteration of *MMP* transcription in relation to musculoskeletal soft tissue injury has generated interest in *MMP* genes per se in Achilles-related pathology in human populations [11, 84]. Indeed, variation within *MMP* genes is associated with a variety of diseases including lung and oesophageal cancer [85, 86], inflammatory bowel disease [87] and invertebral disc degeneration [88]. In terms of musculoskeletal soft tissue injuries, we have recently shown that three common variants within the *MMP3* gene (rs679620, rs591058 and rs650108) are associated with Achilles tendinopathy in a South African Caucasian population [89]. Individuals that are homozygous for the risk alleles at the above loci are at least twice more likely to develop Achilles tendinopathy than individuals that do not have the risk genotypes. Interestingly, the study also found that the G-allele of rs679620 interacted with the T-allele of rs12722 (within the *COL5A1* gene) to further increase risk of tendinopathy [89]. Future work should attempt to establish whether *MMP* sequences variants are associated with other musculoskeletal soft tissue injuries such as ACL.

The ADAMTS are a family of 19 structurally related zinc-containing metalloproteases that catalyse a range of substrates including proteoglycans [90], several procollagens [90], thrombospondin 1 and 2 (ADAMTS 1) [91] and Von Willebrand factor (ADAMTS 13) [92]. A number of growth factors and hormones are known to modulate *ADAMTS* gene expression [92, 93], and strain-induced tendon abnormalities have been reported to influence the expression of *ADAMTS1* in an ovine model [94].

Studies on the expression of a large number of metalloproteases, including all *ADAMTSs*, in normal and degenerative Achilles tendon have revealed a reduction in *ADAMTS 5* expression in painful Achilles compared with controls [95]. The same study also demonstrated that ADAMTS 7 and ADAMTS 13 were repressed in ruptured Achilles compared with healthy controls [95]. Recently the rs4747096 variant within the *ADAMTS14* gene has been shown to associate with various osteoarthritis phenotypes in Caucasians [96]. Whether this variant predisposes to tendon or ligament injury remains unknown. Although it has previously been suggested that *ADAMTS* gene variants should be considered as candidates for musculoskeletal soft tissue injuries [11], no population-based genetic association studies, to date, have emerged in the literature.

Conclusion

To summarise, genetic association studies have shown that sequence variants within three genes (*TNC*, *COL5A1* and *MMP3*) co-segregate with chronic Achilles tendinopathy [29, 30, 58, 89]. The variant within the *TNC* gene also appears to co-segregate with Achilles tendon ruptures [30], while sequence variants within the *COL1A1*, *COL5A1* and *COL12A1* genes have been shown to be associated with cruciate ligament ruptures and/or shoulder dislocations [32, 35, 36]. Functional variants within pyrophosphate metabolism genes (*ANKH* and *TNAP*), on the other hand, are associated with rotator cuff tear arthropathy [97]. Whether these variants are directly involved in the development of these musculoskeletal soft tissue abnormalities or in strong linkage disequilibrium with actual disease-causing loci remains to be established. However, all the genes that contain sequence variants shown to be associated with soft tissue injuries to date encode for proteins directly involved in biological processes within tendons and/or ligaments.

In terms of additional molecular mechanisms that may predispose to musculoskeletal soft tissue injury, recent work has demonstrated that certain CpG sites within the promoters of *MMP3, MMP13* and *ADAMTS 4* genes are 'unsilenced' by DNA hypomethylation in osteoarthritic chondrocytes [98]. It has been postulated that these epigenetic factors may lead to the elevated expression of these proteins in the disease [98]. Accordingly, establishing whether DNA methylation profiles are altered in genes associated with musculoskeletal soft tissue injury would be a future area worthy of research.

In conclusion, using genetic association studies, investigators have started identifying genetic variants which predispose individuals to altered risk of specific acute and overuse musculoskeletal soft tissue injuries. Although it is too early to speculate with respect to specific musculoskeletal soft tissue injuries, the current evidence suggests that different genetic variants may be associated with acute and overuse injuries. Evidence is also starting to emerge that these conditions are polygenic in nature. Research in the area could be accelerated if international consortia, which use high throughput genome-wide association or sequencing technologies on cohorts consisting of larger sample sizes, were established. Prospective follow-up studies are also required to confirm these initial findings, while any proposed molecular mechanism should to be tested using cell culture and other systems. We propose that eventually genetic risk factors will be included in multifactorial models developed to understand the molecular mechanisms that cause musculoskeletal soft tissue injuries. It is conceivable that these models could in the future also be used by clinicians and other professionals to develop personalised training programmes to reduce the risk of injury as well as treatment and rehabilitation regimens for injured athletes.

Acknowledgements

The authors would like to thank Mike Posthumus and Dr Alison September for their useful comments and suggestions after carefully reading the manuscript.

References

1 Ljungqvist A, Schwellnus MP, Bachl N, Collins M, Cook J, Khan KM, Maffulli N, Pitsiladis Y, Riley G, Golspink G, Venter D, Derman EW, Engebretsen L, Volpi P: International Olympic Committee consensus statement: molecular basis of connective tissue and muscle injuries in sport. Clin Sports Med 2008; 27:231–239.

2 Meeuwisse WH: Assessing causation in sport injury: a multifactorial model. Clin J Sport Med 1994;4:166–170.

3 Rees JD, Wilson AM, Wolman RL: Current concepts in the management of tendon disorders. Rheumatology 2006;45:508–521.

4 Jarvinen TA, Kannus P, Maffulli N, Khan KM: Achilles tendon disorders: etiology and epidemiology. Foot Ankle Clin 2005;10:255–266.

5 Kujala UM, Sarna S, Kaprio J: Cumulative incidence of Achilles tendon rupture and tendinopathy in male former elite athletes. Clin J Sport Med 2005; 15:133–135.

6 Marshall SW, Padua D, McGrath M: Incidence of ACL injuries; in Hewett TE, Schultz SJ, Griffin LY (eds): Understanding and Preventing Noncontact ACL Injuries. Champaign, Human Kinetics, 2007, pp 5–30.

7 Gianotti SM, Marshall SW, Hume PA, Bunt L: Incidence of anterior cruciate ligament injury and other knee ligament injuries: a national population-based study. J Sci Med Sport, in press.

8 Parkkari J, Pasanen K, Mattila VM, Kannus P, Rimpela A: The risk for a cruciate ligament injury of the knee in adolescents and young adults: a population-based cohort study of 46,500 people with a 9 year follow-up. Br J Sports Med 2008;42:422–426.

9 Brooks JHM, Fuller CW, Kemp SPT, Reddin DB: Epidemiology of injuries in English professional rugby union. 1. Match injuries. Br J Sports Med 2005;39:757–766.

10 September AV, Mokone GG, Schwellnus MP, Collins M: Genetic risk factors for Achilles tendon injuries. Int Sportmed J 2006;7:201–215.

11 September AV, Schwellnus MP, Collins M: Tendon and ligament injuries: the genetic component. Br J Sports Med 2007;41:241–246.

12 Greenspan DS, Pasquinelli AE: BstUI and DpnII RFLPs at the COL5A1 gene. Hum Mol Genet 1994; 3:385.

13 Frank CB: Ligament structure, physiology and function. J Musculoskelet Neuronal Interact 2004;4:199–201.

14 Hoffmann A, Gross G: Tendon and ligament engineering in the adult organism: mesenchymal stem cells and gene-therapeutic approaches. Int Orthop 2007;31:791–797.

15 Silver FH, Freeman JW, Seehra GP: Collagen self-assembly and the development of tendon mechanical properties. J Biomech 2003;36:1529–1553.

16 Xu Y, Murrell GA: The basic science of tendinopathy. Clin Orthop Relat Res 2008;466:1528–1538.

17 Riley G: The pathogenesis of tendinopathy. A molecular perspective. Rheumatology 2004;43:131–142.

18 Riley G: Tendinopathy – from basic science to treatment. Nat Clin Pract Rheumatol 2008;4:82–89.

19 Kannus P: Structure of the tendon connective tissue. Scand J Med Sci Sports 2000;10:312–320.

20 Hildebrand KA, Frank CB, Hart DA: Gene intervention in ligament and tendon: current status, challenges, future directions. Gene Ther 2004;11:368–378.

21 Ireland D, Harrall R, Curry V, Holloway G, Hackney R, Hazleman B, Riley G: Multiple changes in gene expression in chronic human Achilles tendinopathy. Matrix Biol 2001;20:159–169.

22 Alfredson H, Lorentzon M, Backman S, Backman A, Lerner UH: cDNA-arrays and real-time quantitative PCR techniques in the investigation of chronic Achilles tendinosis. J Orthop Res 2003;21:970–975.

23 Niyibizi C, Kavalkovich K, Yamaji T, Woo SL: Type V collagen is increased during rabbit medial collateral ligament healing. Knee Surg Sports Traumatol Arthrosc 2000;8:281–285.

24 Griffin LY, Albohm MJ, Arendt EA, et al: Understanding and preventing noncontact anterior cruciate ligament injuries: a review of the Hunt Valley II meeting, January 2005. Am J Sports Med 2006;34:1512–1532.

25 Bahr R, Holme I: Risk factors for sports injuries – a methodological approach. Br J Sports Med 2003;37:384–392.

26 Bahr R, Krosshaug T: Understanding injury mechanisms: a key component of preventing injuries in sport. Br J Sports Med 2005;39:324–329.

27 Kannus P, Natri A: Etiology and pathophysiology of tendon ruptures in sports. Scand J Med Sci Sports 1997;7:107–112.

28 Aroen A, Helgo D, Granlund OG, Bahr R: Contralateral tendon rupture risk is increased in individuals with a previous Achilles tendon rupture. Scand J Med Sci Sports 2004;14:30–33.

29 Mokone GG, Gajjar M, September AV, Schwellnus MP, Greenberg J, Noakes TD, Collins M: The Guanine-thymine dinucleotide repeat polymorphism within the tenascin-C gene is associated with Achilles tendon injuries. Am J Sports Med 2005;33:1016–1021.

30 Mokone GG, Schwellnus MP, Noakes TD, Collins M: The COL5A1 gene and Achilles tendon pathology. Scand J Med Sci Sports 2006;16:19–26.

31 Harvie P, Ostlere SJ, Teh J, McNally EG, Clipsham K, Burston BJ, Pollard TC, Carr AJ: Genetic influences in the aetiology of tears of the rotator cuff. Sibling risk of a full-thickness tear. J Bone Joint Surg Br 2004;86:696–700.

32 Khoschnau S, Melhus H, Jacobson A, Rahme H, Bengtsson H, Ribom E, Grundberg E, Mallmin H, Michaelsson K: Type I collagen alpha1 Sp1 polymorphism and the risk of cruciate ligament ruptures or shoulder dislocations. Am J Sports Med 2008; 36:2432–2436.

33 Harner CD, Paulos LE, Greenwald AE, Rosenberg TD, Cooley VC: Detailed analysis of patients with bilateral anterior cruciate ligament injuries. Am J Sports Med 1994;22:37–43.

34 Flynn RK, Pedersen CL, Birmingham TB, Kirkley A, Jackowski D, Fowler PJ: The familial predisposition toward tearing the anterior cruciate ligament: a case control study. Am J Sports Med 2005;33:23–28.

35 Posthumus M, September AV, Keegan M, O'Cuinneagain D, van der Merwe W, Schwellnus MP, Collins M: Genetic risk factors for anterior cruciate ligament ruptures: COL1A1 gene variant. Br J Sports Med 2009;43:352–356.

36 Posthumus M, September AV, O'Cuinneagain D, van der Merwe W, Schwellnus MP, Collins M: The *COL5A1* gene is associated with increased risk of anterior cruciate ligament rupture in females. Am J Sports Med, in press.

37 Collins M, Mokone GG, September AV, van der Merwe L, Schwellnus MP: The *COL5A1* genotype is associated with range of motion measurements. Scand J Med Sci Sports, in press.

38 van der Rest M, Garrone R: Collagen family of proteins. FASEB J 1991;5:2814–2823.

39 Myllyharju J, Kivirikko KI: Collagens and collagen-related diseases. Ann Med 2001;33:7–21.

40 Mann V, Ralston SH: Meta-analysis of COL1A1 Sp1 polymorphism in relation to bone mineral density and osteoporotic fracture. Bone 2003;32:711–717.

41 Lian K, Zmuda JM, Nevitt MC, Lui L, Hochberg MC, Greene D, Li J, Wang J, Lane NE: Type I collagen alpha1 Sp1 transcription factor binding site polymorphism is associated with reduced risk of hip osteoarthritis defined by severe joint space narrowing in elderly women. Arthritis Rheum 2005; 52:1431–1436.

42 Skorupski P, Krol J, Starega J, Adamiak A, Jankiewicz K, Rechberger T: An alpha-1 chain of type I collagen Sp1-binding site polymorphism in women suffering from stress urinary incontinence. Am J Obstet Gynecol 2006;194:346–350.

43 Tilkeridis C, Bei T, Garantziotis S, Stratakis CA: Association of a COL1A1 polymorphism with lumbar disc disease in young military recruits. J Med Genet 2005;42:e44.

44 Speer G, Szenthe P, Kosa JP, Tabak AG, Folhoffer A, Fuszek P, Cseh K, Lakatos P: Myocardial infarction is associated with Spl binding site polymorphism of collagen type 1A1 gene. Acta Cardiol 2006;61:321–325.

45 Mann V, Hobson EE, Li B, Stewart TL, Grant SF, Robins SP, Aspden RM, Ralston SH: A COL1A1 Sp1 binding site polymorphism predisposes to osteoporotic fracture by affecting bone density and quality. J Clin Invest 2001;107:899–907.

46 Posthumus M, September AV, Schwellnus MP, Collins M: Investigation of the Sp1-binding site polymorphism within the COL1A1 gene in participants with Achilles tendon injuries and controls. J Sci Med Sport 2009;12:184–189.

47 Egger G, Liang G, Aparicio A, Jones PA: Epigenetics in human disease and prospects for epigenetic therapy. Nature 2004;429:457–463.

48 Collins M, Posthumus M, Schwellnus MP: The COL1A1 gene and acute soft tissue ruptures. Br J Sports Med, in press.

49 Bray MS, Hagberg JM, Perusse L, Rankinen T, Roth SM, Wolfarth B, Bouchard C: The human gene map for performance and health-related fitness phenotypes: the 2006–2007 update. Med Sci Sports Exerc 2009;41:35–73.

50 Snyder EE, Walts B, Perusse L, Chagnon YC, Weisnagel SJ, Rankinen T, Bouchard C: The human obesity gene map: the 2003 update. Obes Res 2004; 12:369–439.

51 Pearson JV, Huentelman MJ, Halperin RF, Tembe WD, Melquist S, Homer N, Brun M, Szelinger S, Coon KD, Zismann VL, Webster JA, Beach T, Sando SB, Aasly JO, Heun R, Jessen F, Kolsch H, Tsolaki M, Daniilidou M, Reiman EM, Papassotiropoulos A, Hutton ML, Stephan DA, Craig DW: Identification of the genetic basis for complex disorders by use of pooling-based genomewide single-nucleotide-polymorphism association studies. Am J Hum Genet 2007;80:126–139.

52 Attia J, Ioannidis JP, Thakkinstian A, McEvoy M, Scott RJ, Minelli C, Thompson J, Infante-Rivard C, Guyatt G: How to use an article about genetic association. B. Are the results of the study valid? JAMA 2009;301:191–197.

53 Gibson WT: Genetic association studies for complex traits: relevance for the sports medicine practitioner. Br J Sports Med 2009;43:314–316.

54 Birk DE: Type V collagen: heterotypic type I/V collagen interactions in the regulation of fibril assembly. Micron 2001;32:223–237.

55 Chiquet M: Regulation of extracellular matrix gene expression by mechanical stress. Matrix Biol 1999; 18:417–426.

56 Jarvinen TA, Jozsa L, Kannus P, Jarvinen TL, Kvist M, Hurme T, Isola J, Kalimo H, Jarvinen M: Mechanical loading regulates tenascin-C expression in the osteotendinous junction. J Cell Sci 1999;112: 3157–3166.

57 Jarvinen TA, Jozsa L, Kannus P, Jarvinen TL, Hurme T, Kvist M, Pelto-Huikko M, Kalimo H, Jarvinen M: Mechanical loading regulates the expression of tenascin-C in the myotendinous junction and tendon but does not induce de novo synthesis in the skeletal muscle. J Cell Sci 2003;116:857–866.

58 September AV, Cook J, Handley CJ, van der Merwe L, Schwellnus MP, Collins M: Variants within the COL5A1 gene are associated with Achilles tendinopathy in two populations. Br J Sports Med 2009; 43:357–365.

59 Reed CC, Iozzo RV: The role of decorin in collagen fibrillogenesis and skin homeostasis. Glycoconj J 2002;19:249–255.

60 Chakravarti S: Functions of lumican and fibromodulin: lessons from knockout mice. Glycoconj J 2002;19:287–293.

61 Bornstein P, Kyriakides TR, Yang Z, Armstrong LC, Birk DE: Thrombospondin 2 modulates collagen fibrillogenesis and angiogenesis. J Investig Dermatol Symp Proc 2000;5:61–66.

62 Ezura Y, Chakravarti S, Oldberg A, Chervoneva I, Birk DE: Differential expression of lumican and fibromodulin regulate collagen fibrillogenesis in developing mouse tendons. J Cell Biol 2000;151:779–788.

63 Svensson L, Narlid I, Oldberg A: Fibromodulin and lumican bind to the same region on collagen type I fibrils. FEBS Lett 2000;470:178–182.

64 Danielson KG, Baribault H, Holmes DF, Graham H, Kadler KE, Iozzo RV: Targeted disruption of decorin leads to abnormal collagen fibril morphology and skin fragility. J Cell Biol 1997;136:729–743.

65 Nishiyama T, McDonough AM, Bruns RR, Burgeson RE: Type XII and XIV collagens mediate interactions between banded collagen fibers in vitro and may modulate extracellular matrix deformability. J Biol Chem 1994;269:28193–28199.

66 Kania AM, Reichenberger E, Baur ST, Karimbux NY, Taylor RW, Olsen BR, Nishimura I: Structural variation of type XII collagen at its carboxyl-terminal NC1 domain generated by tissue-specific alternative splicing. J Biol Chem 1999;274:22053–22059.

67 Shaw LM, Olsen BR: FACIT collagens: diverse molecular bridges in extracellular matrices. Trends Biochem Sci 1991;16:191–194.

68 Mayne R, Brewton RG: New members of the collagen superfamily. Curr Opin Cell Biol 1993;5:883–890.

69 Olsen BR: New insights into the function of collagens from genetic analysis. Curr Opin Cell Biol 1995;7:720–727.

70 Schuppan D, Cantaluppi MC, Becker J, Veit A, Bunte T, Troyer D, Schuppan F, Schmid M, Ackermann R, Hahn EG: Undulin, an extracellular matrix glycoprotein associated with collagen fibrils. J Biol Chem 1990;265:8823–8832.

71 Keene DR, Lunstrum GP, Morris NP, Stoddard DW, Burgeson RE: Two type XII-like collagens localize to the surface of banded collagen fibrils. J Cell Biol 1991;113:971–978.

72 Zhang X, Schuppan D, Becker J, Reichart P, Gelderblom HR: Distribution of undulin, tenascin, and fibronectin in the human periodontal ligament and cementum: comparative immunoelectron microscopy with ultra-thin cryosections. J Histochem Cytochem 1993;41:245–251.

73 Walchli C, Koch M, Chiquet M, Odermatt BF, Trueb B: Tissue-specific expression of the fibril-associated collagens XII and XIV. J Cell Sci 1994;107:669–681.

74 September AV, Posthumus M, Van der ML, Schwellnus M, Noakes TD, Collins M: The COL12A1 and COL14A1 genes and Achilles tendon injuries. Int J Sports Med 2008;29:257–263.

75 Somerville RP, Oblander SA, Apte SS: Matrix metalloproteinases: old dogs with new tricks. Genome Biol 2003;4:216.

76 Gill SE, Parks WC: Metalloproteinases and their inhibitors: regulators of wound healing. Int J Biochem Cell Biol 2008;40:1334–1347.

77 Pepper MS: Extracellular proteolysis and angiogenesis. Thromb Haemost 2001;86:346–355.

78 Stamenkovic I: Extracellular matrix remodelling: the role of matrix metalloproteinases. J Pathol 2003;200:448–464.

79 Riley GP: Gene expression and matrix turnover in overused and damaged tendons. Scand J Med Sci Sports 2005;15:241–251.

80 Magra M, Maffulli N: Matrix metalloproteases: a role in overuse tendinopathies. Br J Sports Med 2005;39:789–791.

81 Thornton GM, Shao X, Chung M, Sciore P, Boorman RS, Hart DA, Lo IK: Changes in mechanical loading lead to tendon-specific alterations in MMP and TIMP expression: Influence of stress-deprivation and intermittent cyclic hydrostatic compression on rat supraspinatus and Achilles tendons. Br J Sports Med, in press.

82 Pasternak B, Schepull T, Eliasson P, Aspenberg P: Elevation of systemic matrix metalloproteinase-2 and -7 and tissue inhibitor of metalloproteinases-2 in patients with a history of Achilles tendon rupture: pilot study. Br J Sports Med, in press.

83 Higuchi H, Shirakura K, Kimura M, Terauchi M, Shinozaki T, Watanabe H, Takagishi K: Changes in biochemical parameters after anterior cruciate ligament injury. Int Orthop 2006;30:43–47.

84 Magra M, Maffulli N: Genetics: does it play a role in tendinopathy? Clin J Sport Med 2007;17:231–233.

85 Sauter W, Rosenberger A, Beckmann L, Kropp S, Mittelstrass K, Timofeeva M, Wolke G, Steinwachs A, Scheiner D, Meese E, Sybrecht G, Kronenberg F, Dienemann H, Chang-Claude J, Illig T, Wichmann HE, Bickeboller H, Risch A: Matrix metalloproteinase 1 (MMP1) is associated with early-onset lung cancer. Cancer Epidemiol Biomarkers Prev 2008;17:1127–1135.

86 Wu J, Zhang L, Luo H, Zhu Z, Zhang C, Hou Y: Association of matrix metalloproteinases-9 gene polymorphisms with genetic susceptibility to esophageal squamous cell carcinoma. DNA Cell Biol 2008;27:553–557.

87 Meijer MJ, Mieremet-Ooms MA, van Hogezand RA, Lamers CB, Hommes DW, Verspaget HW: Role of matrix metalloproteinase, tissue inhibitor of metalloproteinase and tumor necrosis factor-alpha single nucleotide gene polymorphisms in inflammatory bowel disease. World J Gastroenterol 2007;13:2960–2966.

88 Kalichman L, Hunter DJ: The genetics of intervertebral disc degeneration. Associated genes. Joint Bone Spine 2008;75:388–396.

89 Raleigh SM, Van der ML, Ribbans WJ, Smith RK, Schwellnus MP, Collins M: Variants within the MMP3 gene are associated with Achilles tendinopathy: possible interaction with the COL5A1 gene. Br J Sports Med, in press.

90 Apte SS: A disintegrin-like and metalloprotease (reprolysin type) with thrombospondin type 1 motifs: the ADAMTS family. Int J Biochem Cell Biol 2004;36:981–985.

91 Lee NV, Sato M, Annis DS, Loo JA, Wu L, Mosher DF, Iruela-Arispe ML: ADAMTS1 mediates the release of antiangiogenic polypeptides from TSP1 and 2. EMBO J 2006;25:5270–5283.

92 Jones GC, Riley GP: ADAMTS proteinases: a multidomain, multi-functional family with roles in extracellular matrix turnover and arthritis. Arthritis Res Ther 2005;7:160–169.

93 Porter S, Clark IM, Kevorkian L, Edwards DR: The ADAMTS metalloproteinases. Biochem J 2005;386:15–27.

94 Smith MM, Sakurai G, Smith SM, Young AA, Melrose J, Stewart CM, Appleyard RC, Peterson JL, Gillies RM, Dart AJ, Sonnabend DH, Little CB: Modulation of aggrecan and ADAMTS expression in ovine tendinopathy induced by altered strain. Arthritis Rheum 2008;58:1055–1066.

95 Jones GC, Corps AN, Pennington CJ, Clark IM, Edwards DR, Bradley MM, Hazleman BL, Riley GP: Expression profiling of metalloproteinases and tissue inhibitors of metalloproteinases in normal and degenerate human Achilles tendon. Arthritis Rheum 2006;54:832–842.

96 Rodriguez-Lopez J, Pombo-Suarez M, Loughlin J, Tsezou A, Blanco FJ, Meulenbelt I, Slagboom PE, Valdes AM, Spector TD, Gomez-Reino JJ, Gonzalez A: Association of a nsSNP in ADAMTS14 to some osteoarthritis phenotypes. Osteoarthritis Cartilage 2009;17:321–327.

97 Peach CA, Zhang Y, Dunford JE, Brown MA, Carr AJ: Cuff tear arthropathy: evidence of functional variation in pyrophosphate metabolism genes. Clin Orthop Relat Res 2007;462:67–72.

98 Roach HI, Yamada N, Cheung KS, Tilley S, Clarke NM, Oreffo RO, Kokubun S, Bronner F: Association between the abnormal expression of matrix-degrading enzymes by human osteoarthritic chondrocytes and demethylation of specific CpG sites in the promoter regions. Arthritis Rheum 2005;52:3110–3124.

99 Posthumus M, September AV, O'Cuinneagain D, van der Merwe W, Schwellnus MP, Collins M: The association between the COL12A1 gene and anterior cruciate ligament ruptures. Br J Sports Med, in press.

Prof. Malcolm Collins
UCT/MRC Research Unit for Exercise Science and Sports Medicine
South African Medical Research Council and University of Cape Town
PO Box 115, Newlands 7725 (South Africa)
Tel. +27 21 650 4574, Fax +27 21 686 7530, E-Mail malcolm.collins@uct.ac.za

Innovative Strategies for Treatment of Soft Tissue Injuries in Human and Animal Athletes

Andrea Hoffmann · Gerhard Gross

Department of Molecular Biotechnology, Helmholtz Centre for Infection Research, Braunschweig, Germany

Abstract

Our aim is to review the recent progress in the management of musculoskeletal disorders. We will cover novel therapeutic approaches based on growth factors, gene therapy and cells, including stem cells, which may be combined with each other as appropriate. We focus mainly on the treatment of soft tissue injuries – muscle, cartilage, and tendon/ligament for both human and animal athletes. The need for innovative strategies results from the fact that despite all efforts, the current strategies for cartilage and tendon/ligament still result in the formation of functionally and biomechanically inferior tissues after injury (a phenomenon called 'repair' as opposed to proper 'regeneration'), whereas the outcome for muscle is more favorable. Innovative approaches are urgently needed not only to enhance the outcome of conservative or surgical procedures but also to speed up the healing process from the very long disabling periods, which is of special relevance for athletes.

Copyright © 2009 S. Karger AG, Basel

Injuries of soft tissues of the musculoskeletal system are common and often present as long-time disabling. This situation affects individual persons and even more so professional athletes. Therefore, novel therapeutic strategies need to be developed – in particular based on regenerative medicine and tissue engineering. Regenerative medicine is a multidisciplinary approach to create living, functional substitutes for restoration, maintenance, or improvement of tissue or organ function that has been lost due to injury, disease, age, damage, or congenital defects – either in vitro for subsequent application in vivo, or directly in vivo. This field holds the promise to develop remedies for previously irreparable defects and to solve the problem of organ shortage for transplantation. One important aspect of regenerative medicine is tissue engineering. It includes three main elements: cells, factors or stimuli, and biomaterials. Cells ('grafts') can be autologous or allogeneic and include stem cells where appropriate. Apart from different advantages, choice of donor cells depends on the consideration of disadvantages such as additional trauma during graft harvesting (autograft), infection or immunorejection (allografts). The factors or stimuli are bio(re)active molecules and may be proteins like growth factors and cytokines, drugs, or others. They can be soluble, insoluble, or immobilized. The biomaterials for tissue engineering, which can be derived from natural (like collagen) or synthetic sources (e.g. ceramics, polymers of lactic and glycolic acid), need to be biodegradable and must fulfill several criteria: biocompatibility, capability of integrating molecules or cells, lack of toxicity and immunogenicity, proper biomechanical characteristics, three-dimensional structure, controlled delivery of factors and many more.

Much work within the musculoskeletal system has been done in bone, which is a highly vascularized tissue. Due to the vascularization and medical progress, healing of bone after fracture often proceeds readily. In contrast, soft tissues like cartilage, tendons, and ligaments are poorly vascularized and heal slowly. And their healing ends with the formation of fibrous, scarry tissue which lacks the original flexibility and biomechanical properties.

In this chapter, we will review the recent progress in the management of musculoskeletal disorders focusing specifically on innovative biological methods ('biologicals') including gene therapy for treatment of injuries of muscle, cartilage, and tendon/ligament – for conventional human patients and also for both human and animal athletes.

Wound Healing and Soft Tissue Repair – Current Status

After injury, all tissue healing results from a sequence of defined and consecutive steps: hematoma formation is followed by acute inflammation, then tissue repair by proliferation of fibroblasts and remodeling take place. Ultimately, in many cases less functional tissue is generated in a process called fibrosis or scar formation as opposed to true regeneration which would restore the original functions on a molecular, cellular, functional and biomechanical level. The different stages of healing are not distinct but overlap and are mechanistically intertwined. Their exact timing depends on the type of injury and the severity of damage.

Lesions of the Muscular System

Skeletal muscle is the largest tissue mass in the body, about 40–45% of total body weight, and allows for voluntary force exertion and locomotion. Lesions can be due to direct trauma (e.g. lacerations, contusions, strains, or surgical resection) or have indirect causes like ischemia or neurological dysfunction. Cancer, infectious disease, heart failure, AIDS can produce a body wasting syndrome called cachexia with severe muscle atrophy. Alternatively, muscular dystrophy denotes mostly heritable neuromuscular disorders that likewise cause muscle wasting.

Small muscle injuries like strains can heal completely, but severe injuries typically result in formation of dense scar tissue that impairs muscle function. Muscle injuries in athletes account for 10–55% of sports-related injuries. This frequency is highly variable and depends mainly on the type of sport. Consequently, repair of injured skeletal muscle is of major concern.

Injuries of Cartilage

Cartilage defects can be divided according to their etiology or morphology. As a result of sports injuries, focal injuries typically occur and therefore predominantly affect younger persons. These defects can be subclassified as chondral (cartilage only affected) or osteochondral (defect penetrates into the underlying bone). In older patients, degenerative chondral changes are prevalent. In contrast to bone, cartilage is a tissue with a poor intrinsic capacity for healing which results from poor vascularization (absence of blood vessels and nerve supply), lack of stem cell availability, and/or low cellular turnover, and therefore the efficiency of tissue regeneration is limited. Whereas some spontaneous but transient repair may occur with osteochondral defects since these allow access of stem cells from the vascularized bone to the lesion, chondral defects do not fully regenerate.

Damage to articular cartilage in the knee is a common problem in sports; meniscal tears due to twisting or compression forces are frequent injuries. The menisci are two semilunar fibrocartilaginous structures between the tibia and femur in the knee joint. They function as shock absorbers, joint stabilizers, and joint lubricators. Only tears in the peripheral third of the menisci can heal since this part is vascularized. A special problem exists in the central avascular portion where most tears occur since this never heals and ultimately develops into premature osteoarthritis due to the poor regenerative capacity of articular cartilage which is a prominent example of hyaline cartilage. Sutures, arrows and staples have been used for surgical procedures, but these are only partially successful in alleviating symptoms and do not prevent arthritis development. Apart from total joint arthroplasty which is associated with surgical risks and a finite life span, other techniques developed for therapy include pharmacological intervention, lavage, shaving, laser abrasion, drilling or microfracture of subchondral bone to promote healing, autologous periosteal and perichondral grafting, autologous or allogeneic osteochondral transplantation, and autologous chondrocyte implantation [1]. Autologous and allogeneic grafts for cartilage now are the most common therapeutic approaches. Intense investigations have studied biomaterials on natural or synthetic basis capable of regenerating cartilage defects but so far no good solution has been identified [2], although some recent studies seem quite promising and will be discussed later in the text [3].

Injuries of Tendons and Ligaments

Tendons which connect muscles to bones and ligaments that connect bones to each other are elastic collagenous tissues. They have a similar composition and hierarchical structure of collagen molecules assembling into ever larger units and their strength is related to the number and size of collagen fibrils. Since tendons and ligaments are similar in many aspects we will, for simplicity, only use the term 'tendon' instead of 'tendons and ligaments' in this chapter.

Tendon and ligament lesions can be caused by injury and trauma but also through overuse and ageing. They account for considerable morbidity both in sport and the workplace, and often prove disabling for several months. Orthopedic surgeons see patients with blown-out knees and aching elbows daily. Successful treatment of mostly tendon disorders is difficult and patients often suffer from the symptoms for quite a long time. Moreover, despite remodeling, the biochemical and mechanical properties of healed tendon tissue never match those of intact tendon, which is a major challenge. Tendon naturally heals but – as described above – this process results in repair only, i.e. in the formation of fibrous, scarry tissue which lacks the original flexibility and biomechanical properties and is characterized by a homogenous distribution of smaller than usual diameter collagen fibers and a higher content of collagen type V and decorin and other proteoglycans. Healed tendon therefore shows reduced performance and exhibits a substantial risk for re-injury. Due to this reason, strategies for a veritable tissue regeneration leading to enhanced elasticity and mechanical performance are highly warranted.

Management of tendon injuries currently follows 2 routes: conservative (rest, rehabilitation and pain relief by drugs) or surgical and the healing outcome of both strategies is similar. Experimental and clinical studies have shown that mobilization of the patient is more beneficial than immobilization [4]. As to surgery, conventional treatment comprises sewing torn tendons in place; those that pull free of the bone can be reattached. However, these surgical repairs do not last for a long time – again, novel therapeutic strategies based on biological approaches are needed. In case of extensive damage, grafts (allo- or autografts) have to be used which either bear the risk of infection and rejection (allograft) or donor site morbidity (autograft).

In addition to the difficulties encountered with tendon regeneration per se, the second major obstacle in surgical repair relates to the enthesis, i.e. the site of connection between tendon and bone which is difficult to reproduce. And, such a restored tendon-bone attachment site is the weak link. The attachment sites of tendons to bone are called osteotendinous junction (OTJ or enthesis) that occur in two different forms, according to their histological appearance. The specialized structure of OTJs prevents collagen fiber bending, fraying, shearing, and failure [5]. A fibrocartilaginous (or direct) enthesis is composed of four zones: a dense fibrous connective tissue tendon or ligament zone, uncalcified fibrocartilage, mineralized fibrocartilage, and bone. The outer border of calcification is indicated by a so-called tidemark with basophilic nature, similar to the tidemark found in articular cartilage [6]. The fibrocartilage cells synthesize extracellular matrix that is rich in aggrecan and collagen type II, both of which are typical of articular cartilage. In contrast, a fibrous (or indirect) enthesis lacks the fibrocartilage intermediate zone and is made up of tendon and bone zones only. In any case, full recovery from injury at the junction takes at least 6 months – long for a conventional patient and even longer for a professional sportsman.

Towards Future Innovative Strategies: Animal Models

Despite considerable progress, no optimal solution has been found for the treatment of various sports-related injuries, including muscle injuries, ligament and tendon ruptures, central meniscal tears, cartilage lesions, and delayed bone fracture healing [2]. Therapeutic strategies based on developing novel biological approaches are therefore discussed in the following paragraphs.

Local Delivery of Growth Factors

Growth factors are proteins that stimulate cell division upon binding to their appropriate cell surface receptor(s). One class of growth factors with special relevance to the musculoskeletal system is the transforming growth factor-β family which includes bone morphogenetic proteins (BMPs). BMPs have an important role during bone formation and regeneration and were discovered due to their bone-inducing capacities after implantation at ectopic sites. They are dimeric molecules, most often composed of two identical subunits which transmit their signals predominantly through Smad proteins, a group of related intracellular proteins, from ligand-activated cell surface receptors to the nucleus. Activated R-Smads form heteromeric complexes with the common mediator Smad4 (co-Smad4) that translocate into the nucleus where, in combination with transcription factors, they regulate gene expression. A constitutively active R-Smad – i.e. a Smad protein that is constantly active even in the absence of ligand – can be generated artificially and will be further discussed below.

In general, with growth factor approaches, the major problems are firstly, high doses are needed to be therapeutically efficient which may elicit immune responses, secondly, they easily break down in the body and have very brief half-lives in vivo of only hours whereas a normal healing process takes at least 2 weeks. The use of growth factor proteins to promote healing is severely hindered by the difficulty of ensuring their delivery in minimal yet therapeutically efficient levels to a specific injured site. Various strategies based on scaffolds and biomaterials are therefore being developed (polymers, pumps, heparin, etc.). Alternatively, surfaces of conventional materials may be altered by chemical or physical processes (functionalization by chemical groups, laser structuring of appropriate materials, etc.). For bone repair, approaches including nonspecific and noncovalent absorption of BMP2 on surfaces are providing even better results. Another approach is to chemically modify titanium prostheses or other materials, if necessary including reactive groups that are able to bind growth factors and to release them slowly. We and others could show that such an approach is able to induce biological activity in model cell lines and in vivo as well, see, e.g. [7, 8]. Despite many successful efforts to solve the problem of long-term delivery by strategies like the one just discussed, many researchers may prefer gene-therapeutic strategies (cf. below) or a combination of both, biomaterials and genetically engineered cells or stem cells.

Gene Therapeutic Strategies

In order to transfer genetic material into the target cell(s), specific gene delivery systems called vectors are necessary. They include appropriate regulatory genetic elements and the gene of interest. Such vectors can be based on viruses or on plasmids. Viruses have elaborate systems for transfer of their genetic material into host cells where they replicate themselves and are finally released to infect new cells. Their properties regarding gene delivery can be separated from the pathological consequences of viral replication by removal of selected viral genes. This not only removes the capacity of unwanted and uncontrolled self-replication but also creates space to insert therapeutic genes. Such replication-deficient viral vectors are produced in special cell lines or by transient transfections that provide the viral genes required for replication and packaging into infectious virus. Since they lack the packaging signals that are only attached to the therapeutic gene, the viral genes cannot be packaged into the ensuing virus. Viral vectors have been generated from a wide variety of viruses and therefore, comprise a range of different properties from which they can be selected as applicable. Several systems allow stable and long-term gene expression. Plasmids (nonviral vectors) or naked DNA, on the contrary, are noninfectious and exhibit minimal toxicity. However, they have several drawbacks since their mode of delivery within cationic liposome complexes, as ballistic particles, by calcium phosphate precipitation, or by electroporation is quite unspecific. Transfer efficiency in general is low and expression of the therapeutic gene takes place for only a limited time due to degradation.

Gene transfer can be performed either in or ex vivo, and both strategies need to be considered depending on the individual application and the most suitable route of administration. In ex vivo gene transfer, cells are genetically modified in culture, selected for successfully modified cells by screening or sorting procedures and then given to the recipient either systemically (mainly the blood) or locally (e.g. intra-articularly). The selection processes make ex vivo strategies highly efficient but autologous cell harvesting is an additional burden for the patient and the subsequent cell modifications are time-consuming and expensive. A major advantage but simultaneous limitation of this strategy is that in many cases the treatment could or should be patient-specific to avoid long-term immunosuppression. Examples of therapeutic animal studies in the soft tissues reviewed here will be discussed below in combination with cellular approaches.

In vivo gene transfer, in contrast, means that a vector with the gene of interest can be prepared for the treatment of many individuals, which also reduces the cost of application. In addition, such an approach allows the use of nonintegrating vectors that will lead to relatively short-term gene expression in mitotic (dividing) cells but long-term gene expression in postmitotic cells. In the present context, this applies specifically to muscle fibers. As a disadvantage, no selection and screening of successfully modified cells is possible. Such a somatic gene therapy is legal and would be sufficient for athletes' needs. Local injection at the injury site seems to be the method of choice. The major problem is to ensure that the necessary genes are expressed in the right cells.

We will focus on athletes later on in this chapter. Here, the main target of gene therapy (either for enhanced recovery following injury or degeneration or for improved performance during competitions as 'gene doping') is the musculoskeletal system. Systemic delivery in the musculoskeletal system faces special hurdles since cartilage and tendon lack sufficient blood supply. Despite all efforts, gene therapy is almost exclusively experimental medicine, far from routine applications, and much additional preclinical and clinical studies are necessary to prove efficacy and safety [9]. The misuse of this technique for performance enhancement of athletes bears the risk of serious health problems or side effects including death when used in otherwise healthy people without a clinical need.

Cell- and Stem Cell-Derived Therapies Including Genetically Modified Cells

Due to the limitations of pure growth factor or gene therapy, many labs investigate the usefulness of cells. For regeneration, differentiated cells may be helpful as such or by secreting factors ('trophic effects') that stimulate neighboring or recruited cells to induce a healing response, whereas stem cells may be particularly helpful by differentiating themselves into appropriate cell types. Stem cells have elicited broad interest and excitement due to the promise that either embryonic, adult or induced pluripotent stem cells (derived from somatic cells by overexpression of specific factors) could be used for tissue regeneration for injuries where current therapeutic strategies lack efficiency. Adult stem cells open up the possibility to treat diverse diseases by autologous transplantation (i.e. using the patient's own cells after ex vivo expansion), either through local application or through systemic infusion. Alternatively, since human mesenchymal stem cells (MSCs) are thought to be non- or only faintly immunogenic, even the use of allogenic donor cells might become feasible. They have – due to their possible use in regenerative medicine and tissue engineering – attracted much interest [10]. The transplantation of MSCs into a variety of injured skeletal tissues has been shown to promote healing. In the gene-therapeutic setting, MSCs may be intentionally modified by overexpression of growth or transcription factors to more efficiently conduct their differentiation into appropriate cell types or by secreting factors that stimulate neighboring or recruited cells to induce a healing response. Recently, a subpopulation of human perivascular cells was identified that express both pericyte and MSC markers (one hallmark is expression of the surface antigen CD146, melanoma-associated cell adhesion molecule) in situ, both in vivo and in vitro. The isolated population can be expanded and is clonally multipotent in culture [11]. CD146 positive pericytes can also be serially transplanted to create a hematopoietic environment [12]. These cells seem to be superior to the conventionally isolated MSCs since each single clone is multipotential, and they can be kept in culture for at least 17 weeks or 14 passages [11], whereas conventional MSCs go into senescence already around 6 passages. The use of these stem cells may notably enhance human therapeutic strategies for the musculoskeletal system in the future. Alternatively, the recently described and rapidly expanding field of induced pluripotent stem cells holds great promise of creating patient-specific cells for therapy. Several studies have demonstrated their utility in animal models of human disease like sickle-cell anemia [13] whereas studies in the musculoskeletal system are still missing. However, an inherent tumorigenic potential of these cells has to be closely observed.

As with growth factors, one of the challenges with (stem) cells is to determine the best way to deliver them to their target tissue [14]. To do so, several different biomaterials and scaffolds are in use similar to the ones described for growth factors. One elegant way is to suspend the cells in a viscous fibrin gel which hardens after injection into the body and from which cells can exert their functions. In addition to the mode of delivery, cell transplantation includes more challenges: the way to obtain appropriate cells and to ensure their survival in the recipient.

Exercise and Soft Tissue Regeneration

The musculoskeletal system is particularly subject to mechanical stimuli and loading to properly exert its functions; loss of loading rapidly results in loss of functionality. As a consequence, all components of the musculoskeletal system can be influenced beneficially by exercise and sports and the importance of mechanical stimuli as biological regulators for connective tissue physiology and pathology has been well documented [15–17].

In muscle, exercise may induce satellite cell proliferation and muscle growth [18]. Mechanical stimulation influences metabolic activity, gene expression, and myogenesis as well as fiber alignment [1]. Likewise, many tendon engineering protocols have relied on the application of mechanical load as reviewed, e.g. in reference [19] or [4].

Recent attempts to optimize the functional properties of engineered cartilage have focused on using bioreactors to incorporate mechanical loading environments in vitro to mimic in vivo conditions [1]. In this context, dynamic deformation, hydrostatic pressure, fluid flow, and shear stress were tested, and beneficial effects of mechanical loading on cartilage constructs have been widely reported.

Examples of Novel Treatment Strategies in Animal Models

Muscle
Muscle regeneration occurs early in the healing process. It usually begins 3–5 days after injury, peaks around the 2nd week, and declines thereafter [18]. Various growth factors like insulin-like growth factor-1 (IGF-1), basic fibroblast growth factor (bFGF), or nerve growth factor (NGF) can improve muscle regeneration during these early stages [18]. Local injection of IGF-1, bFGF and – although to a lesser extent – NGF after injury increases the number and size of regenerating myofibers in different mouse injury models. Amongst these, IGF-1 has the greatest beneficial effects since it can increase the efficiency of muscle regeneration and muscle strength in vivo [20]. The so-called Schwarzenegger mouse which is transgenic for IGF-1 presents with enormous muscle generation, i.e. 20–30% greater than in normal mice. In addition, these mice live longer and recover more quickly from injuries [21].

One potential advantage associated with human recombinant growth factors in the treatment of muscle injuries is the ease and safety of the injection procedure that, however, requires a high amount of factors necessary due to rapid clearance by the bloodstream.

The intramuscular implantation of myogenic cells is an approach to develop a therapeutic tool for myopathies, i.e. for diseases with a genetic background [22]. Most studies have been performed on animal models for and patients with Duchenne's muscular dystrophy. For treatment of skeletal muscle, differentiated muscle cells cannot be used because they can neither be properly manipulated nor implanted. Instead, cellular therapy of muscle relies on satellite cells which are the muscle stem cells located on the outer side of muscle fibers, between the basal lamina and plasma membrane. After injury, these cells are triggered to divide and to recapitulate the myogenic events: formation of myoblasts, myocytes, and finally of multinucleated myotubes and muscle fibers. Satellite cells are easily amenable to therapeutic manipulations since they can readily be isolated from muscle biopsies and maintained in culture.

The main factor during intramuscular cell injections to consider is the fact that injected cells fuse mainly with the myofibers damaged by the injection trajectories. As a consequence, cell implantation protocols should include 'high density' injections: 100 cell injections/cm^2 may serve as a guideline [22] but may entail different risks like developing a compartment syndrome in muscles that are enclosed in a rigid osteofacial space. To ensure the long-term survival of myoblast allografts [22], an appropriate immunosuppression would be required. As a consequence, like in the case of gene therapy, such strategies should currently be limited to patients suffering from severe muscular diseases but should not be applied to athletes to speed up a healing process.

The potential therapeutic value of MSCs for muscle repair has been a controversial issue. However, recent studies show that GFP-labeled human MSCs were able to undergo myogenesis upon transplantation. Histology revealed mature muscle characteristics in most GFP-positive myofibers. In addition, some of the cells even seemed to become Pax7-expressing satellite cells [23]. Most recently, the human perivascular cells expressing both pericyte and MSC markers including CD146 have convincingly demonstrated a high potential for muscle fiber generation. This was true not only for cells isolated from skeletal muscle but also for perivascular cells obtained from other donor sites [11].

Readers interested in learning more about therapeutic strategies for muscle injuries might want to see the excellent review by Huard et al. [2].

Cartilage

In contrast to other tissues of the musculoskeletal system, therapeutic approaches using growth factors alone (i.e. in the absence of cells) are not often investigated since they have been even less promising than strategies involving cells. In principle, fully differentiated chondrocytes would be the ideal cell candidate for cartilage repair. Autologous chondrocytes derived from a minor load-bearing area and injected under a periosteal flap can repair full-thickness chondral defects. This procedure (trade name: Carticel™, Genzyme Biosurgery) is the only FDA-approved cell-based therapy for cartilage repair [1]. Apart from being costly and nonconclusively superior to other methods, this autologous technique suffers from potential donor side morbidity, limited availability of chondrocytes, and dedifferentiation of the isolated chondrocytes in monolayer culture. MSCs as a potential alternative to chondrocytes have not proven to be very successful yet. Specifically, their potential to differentiate into chondrocytes in response to transforming growth factor-β is quite limited as a panel of studies have shown.

In the menisci, there are special cells with fibroblast- and chondrocyte-like properties called fibrochondrocytes. Recent evidence has pointed to the potential use of MSCs for meniscal regeneration: GFP-labeled rat MSCs were seeded in fibrin glue and used to treat meniscal defects. These cells were detectable until 8 weeks after transplantation and promoted repair [24]. However, for a successful meniscal therapy, further optimization of regenerative conditions and factors is necessary. Yet another challenge resides in the fact that a functional meniscal tissue must exhibit notable anisotropic mechanical properties resulting from a highly oriented underlying extracellular matrix [1]. In order to generate a scaffold which has controllable, anisotropic properties and mimicks meniscal extracellular matrix fiber alignment to direct fibrochondrocyte orientation, electrospinning technology was used [25]. This technique is based on a biodegradable polymer and results in cartilage with about 20% of the strength of native cartilage but with quite small pores that are difficult for chondrocytes to penetrate [3].

A variety of other natural and synthetic biomaterials have also been investigated. Moutos et al. [26] have created a scaffold which leaves more space for the cells than the electrospinning material. In particular, a different biodegradable material was used in a novel three-dimensional weaving technique which gives compressive, tensile, and sheer strength highly similar to native cartilage, but it degraded too rapidly in vivo. Using another polymer as the basis, the group hopes to go into clinical studies in 2010 [3]. Chiang et al. [27] seeded chondrocytes into gelation microbeads mixed with a collagen type I gel. This enhanced repair in a porcine cartilage defect model, and the chondrocyte phenotype was maintained better than without the carrier. Many more scaffolds and matrices have been under examination [reviewed, e.g., in 1, 28] but so far they have shown variable effects on cell behavior and function.

Viral vectors have been used in several studies to deliver therapeutic genes to chondrocytes in vitro [2]. In addition, these modified cells were applied in collagen gel or organ tissue explants. In vitro delivery of genes for BMP2, IGF-1, and TGF-β to articular chondrocytes has been shown to increase proteoglycan synthesis by the cultured cells [see, e.g., 29, 30]. After retroviral overexpression of BMP-7 in MSCs and cell delivery on a polyglycolic acid scaffold, a rabbit osteochondral articular cartilage defect showed improved healing [31].

In conclusion, there is still a long way to go and develop the optimal strategy and cell source for cartilage repair especially with respect to functional and biomechanical behavior.

Tendon

Within the past years, the American Food and Drug Administration has approved BMP2 and BMP7 for human application, mainly for bone repair. However, BMPs may also prove efficient for cartilage and tendons. BMP12, e.g., was tested for healing of rotator cuff injuries in sheep [32] using two modes of long-term delivery: a paste and a biodegradable collagen sponge to release BMP12. In treated sheep, shoulders were three times stronger (1.8 kN as compared to 0.6 kN) than in sham-operated animals. However, this strength was much less than for a normal, nonruptured shoulder (3.7 kN).

As mentioned in the introduction, one major obstacle in tendon ruptures is the fact that tendon-bone junctions cannot easily be reconstructed to provide secure healing: Especially fibrocartilaginous osteotendinous junctions do not regenerate properly and all attempts for generation of such an enthesis with four zones have failed so far and resulted in fibrous enthesis only [33]. And, an enthesis in the wrong anatomic location entails limited mechanical strength. Therefore, one recent study investigated the effect of rhBMP-2 as compared to noggin, a potent BMP inhibitor. Indeed, BMP2 demonstrated a strong, positive dose-dependent effect on osteointegration at the tendon-bone junction. In contrast, noggin decreased osteointegration. Further studies are necessary to investigate the potential clinical application of enhancing healing and decreasing recovery time with BMPs [34]. A recent study by Hashimoto et al. [35] used application of BMP2 on native rabbit tendon which resulted in ossicle formation including a cartilage region but such an approach requires two surgeries, one to ossify a graft tendon, and a second one to reconstruct the injured site.

Delivery of therapeutic genes has also been investigated for a beneficial use in de novo tendon formation and tendon repair. Several studies were conducted with reporter genes based mainly on adenoviruses in order to assess the overall feasibility of gene transfer into tendon tissues. These studies reveal that direct administration of adenoviruses is possible but may not be the optimal choice for tendon repair since their dense extracellular matrix apparently interferes with efficient infection [10].

In principle, the primary sources of cells for tendon regeneration are currently expected to be tenocytes, MSCs, or tendon stem cells. Tenocytes, derived from tendon progenitor cells, secrete large amounts of collagens which aggregate into basic tendon fibrils. Two major problems with the tenocyte as a therapeutic vehicle, however, arise: Firstly, tendons contain relatively low amounts of cells resulting in limited availability; secondly, they do not proliferate well in culture, dedifferentiate, and lose the characteristic tenocyte morphology [36].

Several studies using MSCs have been conducted and strategies for a successful MSC-dependent repair of tendon tissue have been described. Repair of the Achilles, rotator cuff and patellar tendons has been reported. De novo tendon-like tissue formation was observed after 3 months [37]. In another study, MSC-collagen composites were used for tendon repair of the patellar tendon. In this study, a higher capacity to resist maximum stress than natural repair was reported [38].

Our group could show that a murine MSC-like cell line genetically engineered to express BMP2 and a biologically active Smad8 variant differentiates into tendon-like cells. The engineered cells demonstrated the morphological characteristics and gene expression profile of tendon cells both in vitro and in vivo. In addition, following implantation into an Achilles tendon partial defect, the engineered cells were capable of inducing tendon regeneration [39]. More recently, we could show that by appropriate manipulation of BMP2 and constitutively active Smad8 expression levels the formation of either pure tendon-like tissue or, alternatively, tendon attached to osteotendinous junctions, is possible. Interestingly, C3H10T½ cells generate fibrocartilaginous entheses, whereas human MSCs give fibrous entheses [Shahab-Osterloh et al., unpubl.]. Especially the creation of fibrocartilaginous entheses in vitro would be therapeutically highly desirable.

In 2007, researchers reported that they successfully isolated small numbers of cells with stem cell characteristics from the extracellular matrix of both mouse and human tendons – so called tendon stem cells. Their study showed that adult tendons contain a rare subset of stem cells that can be isolated, grown in the lab, and then used to generate tendon-like tissues at least in mice [40]. Future studies will have to show whether tendon stem cells may be useful in human therapies – one question to answer will be whether these cells can be made available in high enough numbers in a reasonable period of time.

Examples of Therapeutic Strategies in Athletes

Sports injuries usually involve tissues that display a limited capacity for healing. Experience is still limited in human athletes due to concerns about doping [this is discussed by Wells, this vol., pp. 166–175]. However, in the treatment of athletic injuries, growth factor, gene and or cell therapies may induce more rapid healing or result in less side effects from the repair process, such as adhesions and fibrosis. Studies in this field are limited and there is no clear evidence supporting long-term enhanced athletic performance arising from such therapies in animal studies [41]. For instance, expression of growth and differentiation factor-5 in the region of a joint-associated tendon repair improved the range of movement but did not change the (limited) biomechanical properties in a mouse model [42].

Tendon

The anterior cruciate ligament (ACL) consists of a tough band of tissue that connects the thighbone to the shinbone. It stabilizes the knee and enables its motions, including not only stretching but also turning. A torn ACL does not heal well even if surgeons manage to sew the ends back together, which means that athletes cannot compete for several months – and especially for a professional athlete, this is a highly unpleasant situation. Conventional therapy means that often the patellar or hamstring (i.e. semitendinosous plus gracilis) tendon of the patient is used as an autograft to replace the torn ACL. This, however, entails significant donor site morbidity and often proves to be even more disabling than the original defect. Grafting of the central third of the patellar tendon-bone may also lead to patella baja, arthrofibrosis, adhesion to the fat pad, and patellofemoral pain which is hardly helpful for normal patients and even less for sportsmen. Allografts have seen limited use except in revision surgery or for multiple ligamentous injuries [4].

Many studies have been conducted in animals. In particular, there are many similarities between the weight-bearing tendons of the horse and those of the human athlete (e.g. Achilles tendon) in their hierarchical structure and matrix organization, their function and the nature of the injuries sustained. The horse is to be considered as a very special and precious animal athlete suffering often from strain-induced tendon or ligament injuries, resembling consequences of athletic endeavor in humans. Both in horses and in humans, this results in high morbidity and often prevents return to the original performance. Many veterinarians, therefore, are considering therapeutic strategies based on MSCs which, similar to humans, can be preferentially harvested either from equine bone marrow or adipose tissue for treatment of tendon ailments. For horses, MSCs have been applied for the surgical treatment of subchondral bone cysts, bone fracture repair, and cartilage repair [43], and prominently, for tendon injuries. These studies at the same time not only help to cure the horses but also advance our understanding of orthopedic treatment options and enhance translational research into the clinics.

Muscle

Depending on the type of sport, muscle injuries constitute between 10 and 55% of all injuries sustained by athletes [2]. Of relevance for muscle-targeted therapies and gene doping, promoters derived from muscle creatinine kinase, smooth muscle cell α-actin and myosin heavy and light chain have been used in cell culture and animal model studies to achieve muscle-specific expression of recombinant proteins. At present, such strategies should be applied to patients suffering from severe muscular diseases but not to athletes to speed up a healing process because there are many more risks than potential benefits.

Cartilage

As alluded to in the preceding sections of this chapter, the current state of research in cartilage engineering of human patients is not yet entirely promising. However, improvements in culture conditions and scaffold materials, incorporation of bioactive factors, recognition of the role of mechanical stimulation in cartilage formation and novel techniques increasingly contribute to improvement of not only the biological behavior but also the mechanical properties [28]. The 20% of the strength of the native cartilage that has been obtained by several groups is still far from the clinical needs of patients. Such innovative strategies will certainly be of benefit for athletes under the above-mentioned precaution that they will not be abused for doping and give them an illegal advantage over their competitors.

Acknowledgements

The authors gratefully acknowledge the support by the EU integrated project 'GENOSTEM' and by the SFB 599 of the German Research Foundation (DFG) in the past and present years. We apologize to all of our colleagues whose work we could not cite due to space constraints.

References

1 Li WJ, Patterson D, Huang GTJ, Tuan RS: Cell-based therapies for musculoskeletal repair; in Atala A, Lanza R, Thomson J, Nerem R (eds): Principles of Regenerative Medicine. Amsterdam, Elsevier, 2008, pp 888–911.
2 Huard J, Li Y, Peng H, Fu FH: Gene therapy and tissue engineering for sports medicine. J Gene Med 2003;5:93–108.
3 Service RF: Tissue engineering. Coming soon to a knee near you: cartilage like your very own. Science 2008;322:1460–1461.
4 Woo SLY, Almarza AJ, Karaoglu S, Abramowitch SD: Functional tissue engineering of ligament and tendon injuries; in Atala A, Lanza R, Thomson J, Nerem R (eds): Principles of Regenerative Medicine. Amsterdam, Elsevier, 2008, pp 1206–1231.
5 Sharma P, Maffulli N: Biology of tendon injury: healing, modeling and remodeling. J Musculoskelet Neuronal Interact 2006;6:181–190.
6 Benjamin M, Ralphs JR: The cell and developmental biology of tendons and ligaments. Int Rev Cytol 2000;196:85–130.
7 Adden N, Gamble LJ, Castner DG, Hoffmann A, Gross G, Menzel H: Phosphonic acid monolayers for binding of bioactive molecules to titanium surfaces. Langmuir 2006;22:8197–8204.
8 Schmoekel HG, Weber FE, Schense JC, Gratz KW, Schawalder P, Hubbell JA: Bone repair with a form of BMP-2 engineered for incorporation into fibrin cell ingrowth matrices. Biotechnol Bioeng 2005;89: 253–262.
9 Baoutina A, Alexander IE, Rasko JE, Emslie KR: Potential use of gene transfer in athletic performance enhancement. Mol Ther 2007;15:1751–1766.
10 Hoffmann A, Gross G: Tendon and ligament engineering in the adult organism: mesenchymal stem cells and gene-therapeutic approaches. Int Orthop 2007;31:791–797.
11 Crisan M, Yap S, Casteilla L, Chen CW, Corselli M, Park TS, Andriolo G, Sun B, Zheng B, Zhang L, Norotte C, Teng PN, Traas J, Schugar R, Deasy BM, Badylak S, Buhring HJ, Giacobino JP, Lazzari L, Huard J, Peault B: A perivascular origin for mesenchymal stem cells in multiple human organs. Cell Stem Cell 2008;3:301–313.
12 Sacchetti B, Funari A, Michienzi S, Di CS, Piersanti S, Saggio I, Tagliafico E, Ferrari S, Robey PG, Riminucci M, Bianco P: Self-renewing osteoprogenitors in bone marrow sinusoids can organize a hematopoietic microenvironment. Cell 2007;131:324–336.
13 Hanna J, Wernig M, Markoulaki S, Sun CW, Meissner A, Cassady JP, Beard C, Brambrink T, Wu LC, Townes TM, Jaenisch R: Treatment of sickle cell anemia mouse model with iPS cells generated from autologous skin. Science 2007;318:1920–1923.
14 Willyard C: A sporting chance. Nat Med 2008;14: 802–805.
15 Banes AJ: Mechanical strain and the mammalian cell; in Frango JA (ed): Physical Forces and the Mammalian Cell. New York, Academic Press, 1993, pp 81–123.
16 Oddou C, Wendling S, Petite H, Meunier A: Cell mechanotransduction and interactions with biological tissues. Biorheology 2000;37:17–25.
17 Turner CH, Pavalko FM: Mechanotransduction and functional response of the skeleton to physical stress: the mechanisms and mechanics of bone adaptation. J Orthop Sci 1998;3:346–355.
18 Wei S, Huard J: Tissue therapy: implications of regenerative medicine for skeletal muscle; in Atala A, Lanza R, Thomson J, Nerem R (eds): Principles of Regenerative Medicine. Amsterdam, Elsevier, 2008, pp 1232–1247.
19 Hoffmann A, Gross G: Tendon and ligament engineering: from cell biology to in vivo application. Regen Med 2006;1:563–574.
20 Engert JC, Berglund EB, Rosenthal N: Proliferation precedes differentiation in IGF-I-stimulated myogenesis. J Cell Biol 1996;135:431–440.
21 Musaro A, McCullagh K, Paul A, Houghton L, Dobrowolny G, Molinaro M, Barton ER, Sweeney HL, Rosenthal N: Localized Igf-1 transgene expression sustains hypertrophy and regeneration in senescent skeletal muscle. Nat Genet 2001;27:195–200.
22 Skuk D, Tremblay JP: Implantation of myogenic cells in skeletal muscle; in Atala A, Lanza R, Thomson J, Nerem R (eds): Principles of Regenerative Medicine. Amsterdam, Elsevier, 2008, pp 782–793.
23 Dezawa M, Ishikawa H, Itokazu Y, Yoshihara T, Hoshino M, Takeda S, Ide C, Nabeshima Y: Bone marrow stromal cells generate muscle cells and repair muscle degeneration. Science 2005;309:314–317.
24 Izuta Y, Ochi M, Adachi N, Deie M, Yamasaki T, Shinomiya R: Meniscal repair using bone marrow-derived mesenchymal stem cells: experimental study using green fluorescent protein transgenic rats. Knee 2005;12:217–223.

25 Li WJ, Mauck RL, Cooper JA, Yuan X, Tuan RS: Engineering controllable anisotropy in electrospun biodegradable nanofibrous scaffolds for musculoskeletal tissue engineering. J Biomech 2007;40:1686–1693.

26 Moutos FT, Freed LE, Guilak F: A biomimetic three-dimensional woven composite scaffold for functional tissue engineering of cartilage. Nat Mater 2007;6:162–167.

27 Chiang H, Kuo TF, Tsai CC, Lin MC, She BR, Huang YY, Lee HS, Shieh CS, Chen MH, Ramshaw JA, Werkmeister JA, Tuan RS, Jiang CC: Repair of porcine articular cartilage defect with autologous chondrocyte transplantation. J Orthop Res 2005;23:584–593.

28 Shah P, Hillel A, Silverman R, Elisseeff J: Cartilage tissue engineering; in Atala A, Lanza R, Thomson J, Nerem R (eds): Principles of Regenerative Medicine. Amsterdam, Elsevier, 2008, pp 1176–1197.

29 Madry H, Kaul G, Cucchiarini M, Stein U, Zurakowski D, Remberger K, Menger MD, Kohn D, Trippel SB: Enhanced repair of articular cartilage defects in vivo by transplanted chondrocytes overexpressing insulin-like growth factor I (IGF-I). Gene Ther 2005;12:1171–1179.

30 Madry H, Cucchiarini M, Kaul G, Kohn D, Terwilliger EF, Trippel SB: Menisci are efficiently transduced by recombinant adeno-associated virus vectors in vitro and in vivo. Am J Sports Med 2004;32:1860–1865.

31 Mason JM, Grande DA, Barcia M, Grant R, Pergolizzi RG, Breitbart AS: Expression of human bone morphogenic protein 7 in primary rabbit periosteal cells: potential utility in gene therapy for osteochondral repair. Gene Ther 1998;5:1098–1104.

32 Rodeo SA: Biologic augmentation of rotator cuff tendon repair. J Shoulder Elbow Surg 2007;16:S191–S197.

33 Newsham-West R, Nicholson H, Walton M, Milburn P: Long-term morphology of a healing bone-tendon interface: a histological observation in the sheep model. J Anat 2007;210:318–327.

34 Ma CB, Kawamura S, Deng XH, Ying L, Schneidkraut J, Hays P, Rodeo SA: Bone morphogenetic proteins-signaling plays a role in tendon-to-bone healing: a study of rhBMP-2 and noggin. Am J Sports Med 2007;35:597–604.

35 Hashimoto Y, Yoshida G, Toyoda H, Takaoka K: Generation of tendon-to-bone interface 'enthesis' with use of recombinant BMP-2 in a rabbit model. J Orthop Res 2007;25:1415–1424.

36 Bullough R, Finnigan T, Kay A, Maffulli N, Forsyth NR: Tendon repair through stem cell intervention: cellular and molecular approaches. Disabil Rehabil 2008;30:1746–1751.

37 Young RG, Butler DL, Weber W, Caplan AI, Gordon SL, Fink DJ: Use of mesenchymal stem cells in a collagen matrix for Achilles tendon repair. J Orthop Res 1998;16:406–413.

38 Awad HA, Boivin GP, Dressler MR, Smith FN, Young RG, Butler DL: Repair of patellar tendon injuries using a cell-collagen composite. J Orthop Res 2003;21:420–431.

39 Hoffmann A, Pelled G, Turgeman G, Eberle P, Zilberman Y, Shinar H, Keinan-Adamsky K, Winkel A, Shahab S, Navon G, Gross G, Gazit D: Neotendon formation induced by manipulation of the Smad8 signalling pathway in mesenchymal stem cells. J Clin Invest 2006, in press.

40 Bi Y, Ehirchiou D, Kilts TM, Inkson CA, Embree MC, Sonoyama W, Li L, Leet AI, Seo BM, Zhang L, Shi S, Young MF: Identification of tendon stem/progenitor cells and the role of the extracellular matrix in their niche. Nat Med 2007;13:1219–1227.

41 Wells DJ: Gene doping: the hype and the reality. Br J Pharmacol 2008;154:623–631.

42 Basile P, Dadali T, Jacobson J, Hasslund S, Ulrich-Vinther M, Soballe K, Nishio Y, Drissi MH, Langstein HN, Mitten DJ, O'Keefe RJ, Schwarz EM, Awad HA: Freeze-dried tendon allografts as tissue-engineering scaffolds for Gdf5 gene delivery. Mol Ther 2008;16:466–473.

43 Richardson LE, Dudhia J, Clegg PD, Smith R: Stem cells in veterinary medicine – attempts at regenerating equine tendon after injury. Trends Biotechnol 2007;25:409–416.

Dr. Andrea Hoffmann
Helmholtz Centre for Infection Research (HZI), Department of Molecular Biotechnology
Signalling and Gene Regulation, Inhoffenstr. 7
DE–38124 Braunschweig (Germany)
Tel. +49 531 6181 5021, Fax +49 531 6181 5012, E-Mail Andrea.Hoffmann@helmholtz-hzi.de

Gene Doping: Possibilities and Practicalities

Dominic J. Wells

Gene Targeting Group, Department of Cellular and Molecular Neuroscience, Division of Neuroscience and Mental Health, Imperial College London, London, UK

Abstract

Background: Our ever-increasing understanding of the genetic control of cardiovascular and musculoskeletal function together with recent technical improvements in genetic manipulation generates mounting concern over the possibility of such technology being abused by athletes in their quest for improved performance. Genetic manipulation in the context of athletic performance is commonly referred to as gene doping. **Methods:** A review of the literature was performed to identify the genes and methodologies most likely to be used for gene doping and the technologies that might be used to identify such doping. **Results:** A large number of candidate performance-enhancing genes have been identified from animal studies, many of them using transgenic mice. Only a limited number have been shown to be effective following gene transfer into adults. Those that seem most likely to be abused are genes that exert their effects locally and leave little, if any, trace in blood or urine. **Conclusion:** There is currently no evidence that gene doping has yet been undertaken in competitive athletes but the anti-doping authorities will need to remain vigilant in reviewing this rapidly emerging technology. The detection of gene doping involves some different challenges from other agents and a number of promising approaches are currently being explored.

Copyright © 2009 S. Karger AG, Basel

The Concept of Gene Doping

Gene doping is the term applied to the use of genetic manipulation to improve athletic performance. The World Anti-Doping Agency (WADA) prohibits this practice as specifically 'the non-therapeutic use of cells, genes, genetic elements, or the modulation of gene expression, having the capacity to enhance athletic performance'.

Concerns over the possibility of gene doping have been raised following a number of demonstrations of increased running endurance and muscle force after genetic manipulation in rodents. Gene doping offers a number of potential advantages over conventional doping. Firstly, as the protein is made within the body of the treated individual, it will in many cases be indistinguishable from the normal endogenous protein. Secondly, as production of the protein will be steady state, a lower level of continuous expression will likely achieve the same effect as the intermittent high doses given exogenously. Thirdly, in many cases of potential gene doping, the vector and gene product will remain largely in the treated tissue such as the muscle and so will only appear at very trace amounts, if at all, in the routine test samples of blood or urine. All of these properties will make it harder to detect gene doping compared with conventional doping and so increase the allure for athletes who wish to gain an unfair competitive advantage.

Candidate Genes for Improving Athletic Performance

A number of genes have been associated with improved athletic performance, either as the results of identifying individuals with genetic changes associated with enhanced strength or endurance, or through the demonstration of improved performance in genetically modified (GM) animals. Examples of the former include the marked muscle hypertrophy reported in a myostatin null infant [1] and the Finnish cross-country skier Eero Mantyranta who had a natural mutation in his erythropoietin (Epo) receptor gene that increased his haemoglobin levels and so improved endurance [2]. There are many more examples of genetically manipulated animals and some of these are listed in table 1. Of particular interest are those genes that are expressed in the target tissue and will therefore not be present in the circulation. These include transcription factors that regulate the fibre type pattern and oxidative capacity of muscle (e.g. peroxisome proliferator-activated receptor-γ – PPARγ and the coactivators PGC1α and PGC1β).

The reader should be aware that many of the genes identified as offering potential athletic advantage have to date only been reported from studies with transgenic animals. In such transgenic studies, the genetic manipulation occurs in the early embryo such that the modification is carried in every cell of the animal, although depending on the elements controlling expression a gene may only be active in specific tissues or at specific times. In most cases, the genetic modification is expressed during development and the organism may adjust patterns of expression of other genes to accommodate this genetic difference. In contrast, genetic modification in the adult, as might be practiced with gene doping, has not been performed for many candidate genes and so the potential benefits are unclear. For example, modification of muscle phenotype by a transgene may influence motor neuron development to ensure a match but will the same happen following gene transfer into adult muscle? There is one study that has partly addressed this question. Lunde et al. [3] introduced a constitutively active form of a transcription factor, PPAR-δ, which drives oxidative fibre expression, and showed that somatic gene transfer induced similar shifts in fibre-type and oxidative capacity as the transgenic experiment performed by Wang et al. [4] when the lower efficiency gene transfer in the adult was taken into account. It remains to be shown if the same is true for other activators of the same pathway and if such gene transfer changes muscle physiology.

Table 1. A list of examples of genes shown to improve athletic performance in GM animals

Gene	Experimental system	Outcome	Reference
IGF-1	Transgenic mice	Muscle hypertrophy	[32]
IGF-1	AAV-mediated gene transfer in mice	Prevented age-related loss of muscle mass	[33]
IGF-1	AAV-mediated gene transfer in rats	IGF-1 acted synergistically with resistance training to increase muscle power output	[34]
Growth hormone releasing hormone	Plasmid electroporation of muscle in mice, dogs and pigs	Increased muscle mass and other anabolic responses	Reviewed in [35]
PPAR-δ	Transgenic mice	Doubled running endurance compared with wild-type mice	[4]
PPAR-δ	Gene transfer into rat muscle (plasmid electroporation)	Increased IIa and decreased IIb fibres following acute gene transfer	[3]
PGC-1α	Transgenic mice	All muscle type 1 with high fatigue resistance	[36]
PGC-1β	Transgenic mice	High proportion type IIX increased endurance at high work loads	[37]
Phosphoenolpyruvate carboxykinase	Transgenic mice	Substantially enhanced aerobic exercise capacity	[38]
Myostatin deletion	Knockout mice	Muscle hypertrophy	[39]
Myostatin blockade	Hydrodynamic limb vein injection of AAV8 myostatin propeptide (blocks myostatin) in the dog	Muscle hypertrophy in the treated limb	[13]
Epo	Electroporation of plasmid into mouse and non-human primate muscle	Increased hematocrit	[40]
Epo	Intramuscular AAV into non-human primate	Rapamycin-regulated expression gave control of hematocrit	[16]

For further details on these and other genes, please see the excellent review by Baoutina et al. [31].

Improving Recovery from Injury

Apart from improving athletic performance, gene doping can potentially be used to accelerate recovery from injury. The genetic potential has again been demonstrated in GM animals either by transgenesis or following somatic gene transfer. While it would be wrong to deny athletes approved genetic therapies when they become available, there is the potential for athletes to gain a competitive advantage long after the injury has been resolved.

A key feature required for rapid and efficient repair of a bone, joint or tendon defect is an effective blood supply [5]. Neo-vascularisation can be promoted by a number of factors such as vascular endothelial growth factor (VEGF). In addition, other growth factors, such as bone morphogenic protein, activate osteoblasts improving the rate of bone repair. Thus delivery of the DNA encoding such growth factors will enhance repair and so speed recovery from injury. While this would be permitted on therapeutic grounds, it is possible that expression of some of these factors could enhance the strength of bones and tendons, thus giving an advantage to competitors using these techniques, particularly in the strength disciplines like field athletics and weightlifting. Consequently, there is a possibility that such therapies would be used to boost performance in the absence of injury. The current state of the art in orthopaedic gene therapy has recently been reviewed by Evans et al. [6]. They conclude that although the technology is very promising, clinical trials are constrained by safety fears as most orthopaedic conditions are not life threatening unlike many of the other targets for gene therapy. Hence the clinical potential of such methods is unproven.

Muscle repair after injury might also benefit from improving the blood supply with molecules such as VEGF, although a recent review by Yla-Herttuala et al. [7] reveals relatively disappointing results from human clinical trials in cardiovascular disorders to date. IGF-1 overexpression enhances muscle repair in a rodent model of muscle injury [8]. Again both of these treatments for muscle damage also have the potential to increase athletic performance.

Current Practicalities in Gene Delivery

The key three barriers to the successful clinical application of gene therapy are delivery, delivery and delivery. The same problem applies to gene doping. In order to treat conditions such as Duchenne muscular dystrophy (DMD), a lethal muscle wasting disease seen mostly in boys that affects all muscles, a regional or systemic delivery method is necessary for clinical benefit.

Gene transfer can occur either ex vivo or in vivo. In the former case, cells of the individual are grown in culture, genetically modified, selected and then transplanted back into the donor. In the latter, the gene transfer is applied directly to the individual as a local, regional or systemic application. The ex vivo approach requires specialised facilities for cell culture and is labour intensive, but has the advantage that the level of expression of the gene in the GM cells can be assessed. In contrast, the in vivo approach, while much simpler, offers less control over the level of gene transfer and subsequent expression. In both cases, it is necessary to use a system for gene delivery, commonly referred to as a vector.

Heavily modified viruses are used as gene vectors in order to benefit from the natural mechanisms for the delivery of genetic material into cells, an essential part of the viral life cycle. By deleting specific viral genes, replication and consequent cell damage or death is prevented. This also provides space to accommodate the gene(s) of interest. In pre-clinical studies, the most commonly used vectors have been based on retroviruses (including lentiviruses), adenoviruses and adeno-associated viruses (AAVs). Retroviral vectors integrate into the host genome, thus providing a permanent genetic change whereas adenoviral and AAV-based vectors exist as an episome in the nucleus and are lost following cell division. However, for post-mitotic tissues such as muscle the effect of gene transfer is semi-permanent unless the tissue is damaged. Pre-clinical studies in experimental animals ranging from mice to non-human primates have demonstrated that AAV-based vectors can be particularly efficient at gene delivery to muscle and, at least in the mouse, systemic delivery to all muscles is possible [9]. A major current drawback to such vectors is that immune responses to the viral capsid proteins prevent administration of a second dose, although the use of other non-cross-reacting serotypes can circumvent this problem. Alternatively, a regimen of short-term (approximately 6 weeks) immunosuppression at the time of administration may allow repeated treatments. These regimes are currently under intensive investigation to enable the use of such vectors in the clinical treatment of conditions such as DMD.

The alternative to the use of viral vectors is the non-viral system based on purified circular DNA plasmids grown in bacteria. These are very easy to produce and are the basic tool of any molecular biology research project. The delivery of such plasmids to cells generally requires the use of transfection reagents that condense the DNA and encourage uptake by endocytosis into the cells. It is possible to avoid such agents by using physical methods to create transient pores in the cell membrane and so allow entry of the naked plasmid DNA into the cell. These physical methods can be particularly effective in vivo, and techniques such as electroporation [reviewed in 10] and hydrodynamic delivery (reviewed in [11]) can rival the efficiency of viral vectors.

Regardless of the vector system, there are some generic challenges for gene delivery. Local intramuscular injection leads to very limited spread of the vector and so is only suitable where the product will be secreted by the muscle and enter the general circulation (for example Epo or growth hormone). Gene delivery to the whole body does not currently appear to be a realistic option but gene delivery to muscle using perfusion of temporarily isolated limbs (using a tourniquet) has been successful in several animal models using both viral and non-viral vectors [e.g. 12–13]. Expression of the gene can be restricted to muscle by use of a muscle-specific promoter.

A major consideration for the potential gene doper is the ability to control the level of expression of the transferred gene. This will be important as uncontrolled expression may lead to decreased rather than increased performance (see below). As noted above, ex vivo gene transfer offers assessment of the level of gene expression and so one option would be to transfer a given amount of GM cells as a free-standing implant that could if necessary be removed at a later date. Indeed, such an implant can be protected from the action of the immune system and so would not need to utilise cells from the body of the recipient [14].

In the case of in vivo gene transfer, there are drug-inducible systems that would allow athletes to regulate the amount of gene expression. These include the tetracycline, rapamycin and oestrogen regulated systems [e.g. 15–17]. These not only allow control over any harm associated with excessive levels of expression but also provide a method for avoiding doping detection in the case of systemically secreted proteins.

Potential for Harm

Gene doping carries risks associated with the gene delivery and the consequences of gene expression. Large doses of viral vectors risk triggering a massive release of cytokines by the innate immune system and this has proved fatal for one patient in a gene therapy trial [18]. Plasmid DNA can also activate the innate immune system via specific sequences that are recognised as bacterial, and this leads to inflammation and fever. Recent developments in non-viral vector design have largely removed these sequences, which has lead to significantly less interaction with the innate immune system and increased longevity of expression.

Growth hormone and insulin-like growth factor 1 (IGF-1) are both potent mitogens and anti-apoptotic agents and so overexpression following gene transfer carries an increased risk of oncogenesis [reviewed in 19]. Similarly, overexpression of VEGF and other angiogenic factors might lead to increased vascularisation of developing solid cancers and thus could promote tumour growth.

Increasing the number of red blood cells by overexpression of Epo also carries safety risks. An increased haematocrit makes the blood more viscous and increases the load on the heart. This in turn can lead to blockage of the microcirculation including stroke and heart failure. In addition, the production of Epo following gene transfer has been reported as causing autoimmune anaemia in macaques [20, 21].

Detection of Gene Doping

The conventional approach to the detection of doping is to identify a specific banned substance or one of its metabolites using highly sensitive systems such as gas chromatography and high-resolution mass spectrometry. This approach is however limited in that chemical modifications can evade the test, as was seen with the designer anabolic steroid, tetrahydrogestrinone. Consequently, there is increasing interest in methods that can detect the action of a range of doping agents before trying to focus on detecting the specific substance.

One of the challenges facing detection of gene doping is that only body fluid samples are likely to be routinely available. Thus, it will not be possible to take solid tissue samples to look for evidence of genetic modification. Direct detection in blood may be possible in some cases as circulating proteins expressed following gene transfer into ectopic sites show differences in post-translational modification. This was demonstrated for Epo [22] as well as a recombinant form of factor IX [23] following gene transfer into skeletal muscle. It may also be possible to detect minute traces of gene transfer vectors using highly sensitive polymerase chain reaction-based techniques such as the recently described single-copy primer-internal intron-spanning PCR procedure described by Beiter et al. [24]. The evidence to date [reviewed in 25] shows that the window for detection of vector after administration is relatively small, particularly for non-viral vectors. Intramuscular injections leave relatively little trace of vector in the blood and one solution would be to take fine needle aspirates of muscle. With real-time PCR this seems sufficiently sensitive to pick up evidence of genetic manipulation [26]. However, even this is unlikely to be reliable in the case of local intramuscular administration and such an invasive technique seems unlikely to be widely accepted.

Where regulated promoters are used, it may be possible to detect the activating molecule such as tetracycline, rapamycin or tamoxifen. Alternatively, it may be possible to detect evidence of prior administration of viral vectors by looking for evidence of an antibody response to the virus. However, both of these approaches would be subject to legal challenge either under approved medical use or inadvertent exposure to infectious agents.

Another approach is to use indirect techniques to demonstrate potential gene doping by looking for the consequence of the genetic manipulation such as changes in patterns of target gene expression (transcriptomics), proteins (proteomics) or their metabolites (metabalomics). However, these methods would require the establishment of normal standards and some individuals are likely to fall outside of a normal range as a consequence of natural genetic variation. Indeed, the problem of ethnic and individual variation has been highlighted by assessing the metabolism of testosterone [27]. A solution to this problem would be the use of an athlete's endocrinological passport based on the results of repeated tests over time [28] as has been done for "blood passports" for the detection of blood doping [29]. This would provide testers with a lifelong 'biological fingerprint' of competitors to compare drug-test samples against. This passport concept has been in development with WADA since 2006, but it is far from clear if such a system could be widely implemented.

There are several excellent in-depth reviews of specific aspects of the detection of gene doping that should be consulted for further details [e.g. 25, 30].

Current and Future Perspectives

There is no evidence currently that gene doping is being practiced by athletes, but one of the major lessons of the tetrahydrogestrinone scandal and subsequent revelations is how easy it is for athletes to escape detection unless the testing laboratories know what to look for. Thus, quite reasonably there is a concern that gene doping will be used in the near future. The review of current technology presented here suggests that effective gene doping is currently unlikely, but we live in a rapidly changing period of biotechnology and new developments can occur quite suddenly. It is therefore quite right and proper that the authorities such as WADA take this subject seriously and invest in developing potential screening technology, as much to deter athletes from experimenting as to catch those abusing the system.

Acknowledgements

Work in the author's laboratory is primarily focussed on the development of genetic treatments for DMD and over recent years has been sponsored by the Muscular Dystrophy Campaign, the Department of Health, the Medical Research Council, the Big Lottery fund, Parent Project Muscular Dystrophy and the GM Trust.

References

1. Schuelke M, Wagner KR, Stolz LE, Hübner C, Riebel T, Kömen W, Braun T, Tobin JF, Lee SJ: Myostatin mutation associated with gross muscle hypertrophy in a child. N Engl J Med 2004;350:2682–2688.
2. de la Chapelle A, Träskelin AL, Juvonen E: Truncated erythropoietin receptor causes dominantly inherited benign human erythrocytosis. Proc Natl Acad Sci USA 1993;90:4495–4499.
3. Lunde IG, Ekmark M, Rana ZA, Buonanno A, Gundersen K: PPARdelta expression is influenced by muscle activity and induces slow muscle properties in adult rat muscles after somatic gene transfer. J Physiol 2007;582:1277–1287.
4. Wang YX, Zhang CL, Yu RT, Cho HK, Nelson MC, Bayuga-Ocampo CR, Ham J, Kang H, Evans RM: Regulation of muscle fiber type and running endurance by PPARdelta. PLoS Biol 2004;2:e294.
5. Kanczler JM, Oreffo RO: Osteogenesis and angiogenesis: the potential for engineering bone. Eur Cell Mater 2008;15:100–114.
6. Evans CH, Ghivizzani SC, Robbins PD: Orthopedic gene therapy in 2008. Mol Ther 2009;17:231–244.
7. Yla-Herttuala S, Rissanen TT, Vajanto I, Hartikainen J: Vascular endothelial growth factors: biology and current status of clinical applications in cardiovascular medicine. J Am Coll Cardiol 2007;49:1015–1026.
8. Schertzer JD, Lynch GS: Comparative evaluation of IGF-I gene transfer and IGF-I protein administration for enhancing skeletal muscle regeneration after injury. Gene Ther 2006;13:1657–1664.
9. Gregorevic P, Blankinship MJ, Allen JM, Crawford RW, Meuse L, Miller DG, Russell DW, Chamberlain JS: Systemic delivery of genes to striated muscles using adeno-associated viral vectors. Nat Med 2004;10:828–834.
10. Trollet C, Scherman D, Bigey P: Delivery of DNA into muscle for treating systemic diseases: advantages and challenges. Methods Mol Biol 2008;423:199–214.
11. Herweijer H, Wolff JA: Gene therapy progress and prospects: hydrodynamic gene delivery. Gene Ther 2007;14:99–107.
12. Hagstrom JE, Hegge J, Zhang G, Noble M, Budker V, Lewis DL, Herweijer H, Wolff JA: A facile nonviral method for delivering genes and siRNAs to skeletal muscle of mammalian limbs. Mol Ther 2004;10:386–398.

13 Qiao C, Li J, Zheng H, Bogan J, Li J, Yuan Z, Zhang C, Bogan D, Kornegay J, Xiao X: Hydrodynamic limb vein injection of AAV8 canine myostatin propeptide gene in normal dogs enhances muscle growth. Hum Gene Ther 2008, Epub ahead of print.

14 Ponce S, Orive G, Hernández RM, Gascón AR, Canals JM, Muñoz MT, Pedraz JL: In vivo evaluation of EPO-secreting cells immobilized in different alginate-PLL microcapsules. J Control Release 2006; 116:28–34.

15 Nordstrom JL: The antiprogestin-dependent Gene-Switch system for regulated gene therapy. Steroids 2003;68:1085–1094.

16 Rivera VM, Gao GP, Grant RL, Schnell MA, Zoltick PW, Rozamus LW, Clackson T, Wilson JM: Long-term pharmacologically regulated expression of erythropoietin in primates following AAV-mediated gene transfer. Blood 2005;105:1424–1430.

17 Stieger K, Belbellaa B, Le Guiner C, Moullier P, Rolling F: In vivo gene regulation using tetracycline-regulatable systems. Adv Drug Deliv Rev 2009;61: 527–541.

18 Raper SE, Chirmule N, Lee FS, Wivel NA, Bagg A, Gao GP, Wilson JM, Batshaw ML: Fatal systemic inflammatory response syndrome in a ornithine transcarbamylase deficient patient following adenoviral gene transfer. Mol Genet Metab 2003;80:148–158.

19 Perry JK, Emerald BS, Mertani HC, Lobie PE: The oncogenic potential of growth hormone. Growth Horm IGF Res 2006;16:277–289.

20 Chenuaud P, Larcher T, Rabinowitz JE, Provost N, Cherel Y, Casadevall N, Samulski RJ, Moullier P: Autoimmune anemia in macaques following erythropoietin gene therapy. Blood 2004;103:3303–3304.

21 Gao G, Lebherz C, Weiner DJ, Grant R, Calcedo R, McCullough B, Bagg A, Zhang Y, Wilson JM: Erythropoietin gene therapy leads to autoimmune anemia in macaques. Blood 2004;103:3300–3302.

22 Lasne F, Martin L, de Ceaurriz J, Larcher T, Moullier P, Chenuaud P: Genetic Doping' with erythropoietin cDNA in primate muscle is detectable. Mol Ther 2004;10:409–410.

23 Arruda VR, Hagstrom JN, Deitch J, Heiman-Patterson T, Camire RM, Chu K, Fields PA, Herzog RW, Couto LB, Larson PJ, High KA: Posttranslational modifications of recombinant myotube-synthesized human factor IX. Blood 2001;97:130–138.

24 Beiter T, Zimmermann M, Fragasso A, Armeanu S, Lauer UM, Bitzer M, Su H, Young WL, Niess AM, Simon P: Establishing a novel single-copy primer-internal intron-spanning PCR (spiPCR) procedure for the direct detection of gene doping. Exerc Immunol Rev 2008;14:73–85.

25 Baoutina A, Alexander IE, Rasko JE, Emslie KR: Developing strategies for detection of gene doping. J Gene Med 2008;10:3–20.

26 Guescini M, Fatone C, Stocchi L, Guidi C, Potenza L, Ditroilo M, Ranchelli A, Di Loreto C, Sisti D, De Feo P, Stocchi V: Fine needle aspiration coupled with real-time PCR: a painless methodology to study adaptive functional changes in skeletal muscle. Nutr Metab Cardiovasc Dis 2007;17:383–393.

27 Strahm E, Sottas PE, Schweizer C, Saugy M, Dvorak J, Saudan C: Steroid profiles of professional soccer players: an international comparative study. Br J Sports Med 2009, Epub ahead of print.

28 Sottas PE, Saudan C, Schweizer C, Baume N, Mangin P, Saugy M: From population- to subject-based limits of T/E ratio to detect testosterone abuse in elite sports. Forensic Sci Int 2008;174:166–172.

29 Robinson N, Sottas PE, Mangin P, Saugy M: Bayesian detection of abnormal hematological values to introduce a no-start rule for heterogeneous populations of athletes. Haematologica 2007;92: 1143–1144.

30 Minunni M, Scarano S, Mascini M: Affinity-based biosensors as promising tools for gene doping detection. Trends Biotechnol 2008;26:236–243.

31 Baoutina A, Alexander IE, Rasko JE, Emslie KR: Potential use of gene transfer in athletic performance enhancement. Mol Ther 2007;15:1751–1766.

32 Musaro A, McCullagh K, Paul A, Houghton L, Dobrowolny G, Molinaro M, Barton ER, Sweeney HL, Rosenthal N: Localized IGF-1 transgene expression sustains hypertrophy and regeneration in senescent skeletal muscle. Nat Genet 2001;27:195–200.

33 Barton-Davis ER, Shoturma DI, Musaro A, Rosenthal N, Sweeney HL: Viral mediated expression of insulin-like growth factor I blocks the aging-related loss of skeletal muscle function. Proc Natl Acad Sci USA 1998;95:15603–15607.

34 Lee S, Barton ER, Sweeney HL, Farrar RP: Viral expression of insulin-like growth factor-I enhances muscle hypertrophy in resistance-trained rats. J Appl Physiol 2004;96:1097–1104.

35 Khan AS, Brown PA, Draghia-Akli R: Plasmid-based growth hormone-releasing hormone supplementation and its applications. Curr Opin Mol Ther 2005;7:306–316.

36 Lin J, Wu H, Tarr PT, Zhang CY, Wu Z, Boss O, Michael LF, Puigserver P, Isotani E, Olson EN, Lowell BB, Bassel-Duby R, Spiegelman BM: Transcriptional co-activator PGC-1 alpha drives the formation of slow-twitch muscle fibres. Nature 2002;418:797–801.

37 Arany Z, Lebrasseur N, Morris C, Smith E, Yang W, Ma Y, Chin S, Spiegelman BM: The transcriptional coactivator PGC-1beta drives the formation of oxidative type IIX fibers in skeletal muscle. Cell Metab 2007;5:35–46.

38 Hakimi P, Yang J, Casadesus G, Massillon D, Tolentino-Silva F, Nye CK, Cabrera ME, Hagen DR, Utter CB, Baghdy Y, Johnson DH, Wilson DL, Kirwan JP, Kalhan SC, Hanson RW: Overexpression of the cytosolic form of phosphoenolpyruvate carboxykinase (GTP) in skeletal muscle repatterns energy metabolism in the mouse. J Biol Chem 2007; 282:32844–32855.

39 McPherron AC, Lawler AM, Lee SJ: Regulation of skeletal muscle mass in mice by a new TGF-beta superfamily member. Nature 1997;387:83–90.

40 Fattori E, Cappelletti M, Zampaglione I, Mennuni C, Calvaruso F, Arcuri M, Rizzuto G, Costa P, Perretta G, Ciliberto G, La Monica N: Gene electrotransfer of an improved erythropoietin plasmid in mice and non-human primates. J Gene Med 2005; 7:228–236.

Prof. Dominic J. Wells
Department of Cellular and Molecular Neuroscience, Imperial College London
Hammersmith Hospital Campus, 160 Du Cane Road
London W12 0NN (UK)
Tel. +44 207 594 7024, Fax +44 207 594 7025, E-Mail d.wells@imperial.ac.uk

Genetic Testing of Athletes

Alun G. Williams[a] · Henning Wackerhage[b]

[a]Department of Exercise and Sport Science, Manchester Metropolitan University, Alsager, and [b]School of Medical Sciences, University of Aberdeen, Aberdeen, UK

Abstract

This chapter addresses the potential use of genetic tests to predict performance and/or risk of exercise-related injury or illness. Various people may wish to conduct a sport-related genetic test on themselves, or on another person, for a variety of reasons. For example, an individual may seek personal genetic information to assist with choosing their own sporting participation or career. A sports coach may wish to test young team members to assist in selection for a professional career or to individualise training. A physician may want to predict risk of injury or illness in an athlete and advise regarding selection or preventative measures. An insurance company may seek to estimate risk of career-threatening injury or illness to an athlete based partly on genetic information. Despite the commercial availability of some genetic tests today, the evidence currently available suggests that few of these or similar scenarios are scientifically justified – the genetic tests available at the moment are simply not powerful enough to inform important decisions in sport. Furthermore, there are many challenging ethical issues to be addressed regarding genetic testing of athletes. The imposition of genetic tests on individuals by third parties, and particularly the imposition of genetic tests on young people, is potentially susceptible to abuse. There should be considerable further debate regarding these issues so that the tests already available, and the more powerful ones that are likely to emerge as knowledge and genetic technology improve, are only used in acceptable ways.

Copyright © 2009 S. Karger AG, Basel

The focus of this chapter is on the potential use of genetic tests to predict performance and/or risk of exercise-related injury or illness. Some earlier chapters of this book, including chapters by Yang et al. [pp. 88–101] (regarding the ACTN3 R577X polymorphism), Pomeroy et al. [pp. 110–135] (exercise, health and genes) and Collins and Raleigh [pp. 136–149] (regarding soft tissue injuries), have already alluded to the notion of testing athletes or other individuals for sport- and exercise-related traits. However, this chapter will deliberately focus on both the scientific and ethical issues regarding the validity, utility, practicality, and ethics of genetic testing of athletes at the current time.

There are already several commercial operations that offer 'mail order' or 'direct-to-consumer' genetic testing of polymorphisms apparently associated with physical performance capacity or exercise-related health. We believe the first commercial operation began in Australia in 2004 and is marketed as the 'ACTN3 Sports Performance Test^{TM}' [1]. Other commercial operations we are aware of are currently (January 2009) available at 'atlasgene.com', 'mycellfdnafitness.com' and 'cygenelabs.com', some of which claim to test multiple genotypes. The availability of such tests for virtually anyone to access is certainly an interesting development and probably heralds a new era of genetic testing for all manner of individual characteristics. Such tests may become more widely requested because recent legislation in the US ensures that consumers do not have to fear insurance or employment discrimination on the basis of genetic test results [2]. However, significant debate is ongoing regarding the science that is purported to underpin such tests. In addition, there are wider ethical concerns regarding issues such as the confidentiality of data, the need for counselling when interpreting personal genetic data, the use of genetic data in assessing insurance risk, etc. We will address such issues in this chapter.

Is There a Fundamental Difference between Genetic Tests and Non-Genetic Tests?

The film 'GATTACA' (a science fiction film directed by Andrew Niccol about a society where genetic testing is used to determine social class, careers and partners of its citizens) paints a worst-case scenario about discrimination based on the results of genetic testing. 'GATTACA' highlights the concerns that some people have about the potential misuse of genetic testing at a future time when meaningful genetic tests for intellectual or physical ability are available.

Discrimination on the basis of partially inherited factors is already common in sport today. For example, when strength is measured to identify children that have a talent for weightlifting or to select the national weightlifting team. So does it matter whether talent is identified by genetic or by non-genetic tests? Is there a fundamental difference?

To give an example, one could rightly ask 'what is the difference between a performance test that measures a variable that predicts ~2% of an individual's muscle power as opposed to the commercially available ACTN3 R577X genetic test which measures something very similar?' The term 'genetic exceptionalism' is sometimes used in this debate meaning that genetic tests are special and thus require specific legislation that is different than that for other biological tests [3]. In their analysis, Green and Botkin [3] focus on medical tests and conclude that 'no clear, significant distinctions between genetic and non-genetic tests justify a different approach [...]' although they state that genetic tests may lead to stigmatisation, family discord and psychological stress, but much of this also applies to some non-genetic tests such as HIV tests.

The working party that developed the BASES position stand on 'Genetic Research and Testing in Sport and Exercise Science' [4] identified two fundamental differences between genetic and non-genetic (performance) tests, and we add a third in this chapter:

1   The information gained from a genetic test does not change with age, whereas the information derived from a traditional performance test, and consequently the predictive quality of that test, does change with age.
2   Not all the predictions that can be made with a genetic test may be known at the time when the genetic test is conducted.
3   Genetic tests have more implications for relatives and partners than other tests.

These three points are now discussed in more detail:

1   The genome of an individual and the variations therein are constant throughout life (there are exceptions not relevant for this argument). Consequently, genetic tests can be performed as soon as the DNA of an individual can be obtained – even before birth. This is fundamentally different from other tests such as a lactate test where the information derived from the test, and consequently its predictive quality (for example, the ability to predict Marathon running ability), depend very much on the age when this test is performed. This difference is important because it opens up possibilities for the misuse of genetic tests, especially when used before birth or in minors. One worst-case scenario could be that these tests are used to select embryos on the basis of sporting, intellectual or other abilities (a new eugenics). Also, with young children such tests could be used to make decisions for or against a sporting career.
2   Both genetic and non-genetic performance tests may later be linked to other variables such as disease. We think that this risk is higher for genetic tests than for other performance tests. However, common performance-related variables have also been linked to disease: For example, muscle strength has recently been inversely associated with mortality [5], which means a hitherto unanticipated higher risk of death for individuals who had previously performed poorly in a strength test. However, our opinion is that the risk of an unanticipated disease link is probably higher for genetic tests than for other biomedical tests. One past example is a polymorphism for the apolipoprotein E gene (which encodes a protein important for lipid transport). This gene occurs in three polymorphic forms: APO E2, E3 and E4. The APO E4 variant was initially shown to be associated with limited differences in lipid profile but only later with late-onset familial Alzheimer's disease [6]. Thus, individuals who had previously been tested for a genetic variant that was associated with their lipid profile now retrospectively had information about their likelihood of suffering from the more severe Alzheimer's. How can we deal with the risk of unanticipated future links of genetic test results to severe disease? A pragmatic solution would be to make mandatory genetic counselling necessary so that individuals are aware of this risk before deciding whether to proceed or not with a genetic test.
3   A third difference is that genetic tests have more direct implications for relatives than other tests. As Green and Botkin [3] point out, this is also the case for some non-genetic tests. For example, a positive HIV test result may not only be devastating for the individual but has also far-reaching implications for relatives, current

and previous partners. However, genetic tests are always also predictive for close relatives and this should be taken into account when performing genetic tests. Again, mandatory genetic counselling could be useful to at least ensure that individuals understand the implications of genetic test results for their relatives.

The first difference we have highlighted in particular, and its resultant consequences such as the potential selection of embryos, warrant the use of the term 'genetic exceptionalism'. Also, scientists for now at least must accept that 'people see genetic information as special' [7], even if in reality the information obtained by genetic and non-genetic tests can in some instances be very similar.

How Powerful Are the Genetic Tests That Currently Exist?

By the end of 2007, there had been over 200 genetic variations associated with performance and health-related fitness phenotypes [8] and that number continues to increase. In many cases, there has only been a single positive association with a relevant phenotype and consistent replication of the reported associations is obviously needed to increase confidence in the associations.

We are aware of just two rare mutations that each appears to have a powerful 'positive' influence on exercise-related phenotypes (although many rare mutations exist that have powerful 'negative' influences on exercise-related phenotypes). One rare mutation exists in the family of the Finnish cross-country skier Eero Antero Mäntyranta, where over 200 g of haemoglobin per litre of blood and extremely high haematocrit values >60% were reported [9]. EPO levels were normal but the mutation enhanced sensitivity to EPO. The responsible mutation in the EPO receptor gene was subsequently identified [10] as a premature stop codon that shortens the protein by 70 amino acids. A more active EPO receptor protein is the result, and is likely to have contributed to Mäntyranta winning three gold and four other medals at the Olympic games during the 1960s. Haematocrit in excess of 50% would attract investigation from the anti-doping authorities, although the World Anti-Doping Agency (WADA) 'Athlete Passport' concept combined with DNA analysis may allow an athlete competing today with such a mutation to demonstrate that erythropoiesis-stimulating agents had not been used.

A second naturally occurring rare mutation, this time with profound effects on muscle growth, has been observed in the myostatin gene (MSTN [11]). The authors reported the existence of a boy who was 'extraordinarily muscular, with protruding muscles in his thighs and upper arm'. The boy was homozygous for an extremely rare G to A intronic mutation in the myostatin gene and his muscle tissue showed no evidence of functional myostatin protein. This observation in a human parallels the deliberate knockout of the myostatin gene in mice, which has resulted in pronounced skeletal muscle fibre growth [12]. However, it is not certain that health status (e.g. cardiac function) will not be compromised by the absence of functional myostatin protein, and there may be other detrimental changes such as changes in tendon properties [13]. Nevertheless, one possible kind of genetic test for physical performance could be to screen athletes for rare mutations such as those described in the EPOR and MSTN genes. The rarity and heterogeneity of such powerful mutations means that they would need to be identified using sequencing technology that allows base-by-base examination of relevant genes and gene portions, but even then would probably have a very low success or 'hit' rate. Nevertheless, future identification of an athlete with a particular rare, powerful mutation might help to explain the physiological capacity of that individual, provide clues regarding the type of events in which they may excel and even provide information that may be useful in the individualisation of his/her training.

In contrast to rare, powerful mutations, there are many common polymorphisms that are usually less powerfully associated with performance-related phenotypes. Here, we will consider as examples just the two potentially performance-related polymorphisms that have received the most research attention and one interesting polymorphism potentially related to injury risk.

As discussed in the chapter by Woods [pp. 72–87], a common insertion/deletion (I/D) polymorphism at the ACE locus is associated with circulating and tissue ACE activity, and associations of this ACE polymorphism with elite athletic status and exercise-related phenotypes (e.g. [14, 15]) have suggested the I-allele is associated with endurance phenotypes and the D-allele is associated with strength-related phenotypes. The polymorphism has also been associated with muscle fibre type proportion [16] which could go some way to explain these associations. However, other research groups have reported little or no association between the ACE I/D polymorphism and human physical performance (e.g. [17]). Our considered view is that the ACE I/D polymorphism is probably associated with some aspects of physical performance (e.g. physical capability at altitude) in some populations (e.g. well-trained Caucasians), but that the influence of this single polymorphism is probably modest. The technology required for screening athletes for the ACE I/D polymorphism (and most other common polymorphisms) is relatively commonplace (e.g. PCR and gel electrophoresis, or real-time PCR). However, the few percent (at best) of interindividual variability in a phenotype that may be attributable to ACE genotype is not sufficient for a genetic test at the ACE I/D locus to inform an individual (or his/her coach, for example) about future performance potential, which training mode may be most suitable, etc.

As discussed in the chapter by Yang et al. [pp. 88–101], a common polymorphism exists in the ACTN3 gene in which the element encoding arginine (R) may alternatively be a stop codon (X) that produces a truncated protein molecule with no known function. Studies of elite Australian [18] and Finnish [19] sprint/power athletes found none were of XX genotype. An ACTN3 knockout mouse displaying characteristics summarised as a fast-to-slow muscle phenotype change helps to explain mechanistic links between ACTN3 genotype and physical performance phenotype [20]. These data suggest it is virtually impossible to become an elite sprint or power athlete without ACTN3, perhaps suggesting a powerful influence of the ACTN3 common polymorphism. However, in the general population it appears only ~2% of interindividual variability in muscle strength and sprinting speed are related to ACTN3 genotype [21, 22], and this seems somewhat at odds with the striking absence of the XX genotype in most elite sprint/power athletes. This paradox may be analogous with a classic study demonstrating that while maximal rate of oxygen uptake may have good statistical power in predicting endurance running performance amongst a wide range of running abilities, when a narrow (elite) range of performance is considered it is then running economy, a different phenotype, that becomes the dominant factor influencing performance [23]. Perhaps ACTN3 genotype (and thus the presence/absence of α-actinin-3 protein in muscle) is equivalent to running economy in that analogy, in that only amongst highly trained athletes does the practical significance of ACTN3 genotype increase to a meaningful extent. Nevertheless, even then, ACTN3 XX genotype does not seem to completely preclude success in power events [24]. The practical use of an ACTN3 genotype test to inform an individual (or his/her coach, for example) about future performance potential is thus still not clear, although of all common polymorphisms ACTN3 currently shows the most potential in this regard.

As discussed in the chapter by Collins and Raleigh [pp. 136–149], a GT dinucleotide repeat polymorphism in the tenascin-C (TNC) gene, which codes for an extracellular matrix protein, has been associated with the risk of Achilles tendon injury [25]. As in much research of genotype-phenotype interactions that is exploratory, it remains possible that the TNC polymorphism studied is not functional with respect to tendon injuries but merely in linkage disequilibrium with a different, functional polymorphism, perhaps within an adjacent gene. However, the logic for TNC and the protein for which it encodes being functional is strong. But regardless of the functionality, the potential exists to genotype athletes for the TNC polymorphism in an attempt to provide a basis for objective individualisation of training to minimise the risk of Achilles tendon injury or introduce other preventative measures. More research is needed to quantify the power of this polymorphism in predicting injury risk in specific sports, groups of individuals with specific geographic ancestry, etc., but the data are promising.

It is our view that while ACTN3 and TNC genotypes show promise, a large monogenic influence of a common polymorphism on physical performance or injury phenotypes is extremely unlikely. Future genetic testing of athletes will therefore need to account for a greater proportion of the genetic influence on interindividual variability via consideration of genetic variation at a number of loci simultaneously. The principle behind this polygenic approach has been demonstrated recently [26], although clearly more research is needed before the application of genetic algorithms to athletes can proceed with a valid scientific basis. Major decision-making regarding the careers or training/medical environment of athletes, based largely on the genetic testing of those athletes, is not scientifically justified until more research is conducted and therefore would be unethical at the present time. Increased predictive power of genetic testing in athletes will be dependent upon the quality of the data that emerge from the future research [Trent and Yu, this vol., pp. 187–195] that will itself be dependent on the application of more advanced technology such as cheap, fast, whole-genome sequencing.

Who Should Be Able to Request Genetic Tests?

In the sport and exercise context, various people may request genetic tests for different reasons. With an accumulating number of available genetic tests, a wide range of genetic testing scenarios seems inevitable in the sport and exercise context. The five examples below show different individuals (athletes, parents, coaches, physicians) requesting genetic tests either for themselves or for others for a variety of purposes:

1 Genetic performance testing. Parents might request genetic tests for genes that determine adult peak maximal oxygen uptake for their child in order to decide whether to send their offspring to a special sports school.
2 Genetic testing for personalised training programmes. A coach might request genetic tests for trainability polymorphisms in order to decide whether weightlifters should take part in a high volume or high intensity strength training programme [for a more detailed debate, see 27].
3 Genetic sports disease testing. Sports associations and their physicians might use genetic tests to screen for variants associated with sudden death in order to identify individuals that are at high risk and to exclude such individuals from competing. A similar but non-genetic screening programme is performed in Italy [28] and has reduced sudden deaths but has also led to the exclusion of elite athletes from competitions.
4 Genetic sports injury testing. An individual could request a 'direct-to-consumer' genetic test [29] for tendon gene variants to predict the frequency of injury in order to decide whether to embark on a career as a professional football player.
5 Genetic insurance risk testing. An insurance company might request genetic tests for injury genes in order to determine insurance premium for a professional football player.

At the moment, most people would probably accept that adults should not be prohibited from requesting genetic tests for performance or sport and health related traits for themselves, in order to assist making life choices. We recommend, however, even in the case of 'mail order' or 'direct-to-consumer' tests [29] such as the commercially available ACTN3 R577X test [1] that these tests are accompanied by some genetic counselling. This could include counselling about potential unanticipated associations with disease, implications for relatives and other information which has been summarised in a review on 'ideal' genetic counselling [30]. The other recommendation is that genetic tests should, for the time being at least, only be allowed to be requested by mature individuals (in the sense of mental capacity, in specific relation to the issue of genetic testing in sport) who understand the implications of a test.

Serious ethical concerns arise if genetic tests for an individual are requested by others. In the sport context this could be physicians, parents or coaches, and such tests could be used to discriminate in favour of or against athletes on the basis of performance-related or health-related genetic information. Some of these concerns are highlighted by a real case [31]. Eddy Curry is a professional NBA basketball player who missed games due to cardiovascular symptoms. His team (the Chicago Bulls) requested a genetic test for him based on the advice of a cardiologist. The athlete refused to do the test and was sold to the New York Knicks who did not request such a test. This is an example of an employer trying to force an employee to do a genetic test in order to then discriminate for or against the employee based on the test result. In this case, one can see the team's perspective as an employer that has a duty of care that might want to prepare itself for an acute cardiovascular problem, yet on the other hand one can understand the athlete's desire not to know or have others know about a serious genetic condition.

In an increasing number of countries, there is a legal framework that deals with genetic testing and resultant discrimination. In 2008, US Congress voted in favour of the 'Genetic Information Nondiscrimination Act' (GINA) to 'prohibit discrimination on the basis of genetic information with respect to health insurance and employment' in the USA [2]. In the sporting context, the WADA states in its 'Stockholm Declaration' [32]: 'The use of genetic information to select for or discriminate against athletes should be strongly discouraged', although the precise rationale for that statement is not given. It would be interesting to see whether the Eddy Curry case would be covered by GINA, although WADA also states that '[discouraging genetic discrimination] does not apply to legitimate medical screening'. However, both documents would prevent or strongly discourage professional football clubs, for example, from performing ACTN3 R577X tests on their players and using such information to discriminate against players. The international legal framework and related ethical issues for genetic testing have recently been reviewed [33].

The most serious ethical concern arises from the major fundamental difference between genetic and non-genetic testing which is that the DNA does not change throughout life and that therefore prenatal genetic tests could be performed to select an embryo with the 'best sport genotype' or to abort an embryo with the 'wrong sport genotype'. The legal prevention of such testing is important because there is evidence that some parents use pre-natal information to decide about whether or not to abort a pregnancy. In India and China, non-genetic ultrasound scans have been used to determine the sex of an embryo and this practice has greatly contributed to the estimated 80 million 'missing' females [34], a shocking statistic. Similarly, it may be that some genetic tests (not limited to genetic sex tests), including genetic tests of sporting potential, would be used by some parents during pre-implantation genetic diagnosis or to decide whether or not to abort an embryo.

Conclusions

The focus of this chapter has been on the potential use of genetic tests to predict performance and/or risk of exercise-related injury or illness. Various people may wish to conduct a sport-related genetic test on themselves, or on another person, for a variety of reasons. An individual may seek personal genetic information to assist him/her with their own sporting participation and career, by identifying the most suitable type of sport. A sports coach may wish to test the members of a youth team to assist in selection for a professional career or to individualise training. A physician may want to predict risk of injury or illness in an athlete and advise a coach regarding selection or preventative measures. An insurance company may seek to estimate risk of career-threatening injury or illness to an athlete based partly on genetic information. Despite the commercial availability of genetic tests today, the evidence available at present suggests that few, and probably none, of these or similar scenarios are scientifically justified – the genetic tests available at the moment are simply not powerful enough to provide valid data on which to base important decisions in sport. Furthermore, there are many challenging ethical issues regarding genetic testing of athletes. The imposition of genetic tests on individuals by third parties, and particularly the imposition of genetic tests on young people, is potentially susceptible to abuse. We propose that, for now at least, an individual should only be able to request a genetic test of their *own* DNA (i.e. not of another individual/employee/child), and only if he/she can demonstrate beforehand an understanding of the possible implications of the test result. We also recommend that an appropriate standard of genetic counselling is always provided. It is critical that these issues continue to be debated widely so that the tests already available, and the more powerful ones that are likely to emerge following improvements in knowledge and genetic technology, are only used in acceptable ways.

References

1 Genetic Technologies Limited: Human DNA Testing: Sports Performance. Genetic Technologies Limited, 2008. http://www.gtg.com.au/HumanDNATesting/index.asp?menuid = 070.110; retrieved 11 December 2008.
2 National Human Genome Research Institute: Genetic Information Nondiscrimination Act: 2007–2008. National Human Genome Research Institute, 2008. http://www.genome.gov/24519851; retrieved 7 January 2009.
3 Green MJ, Botkin JR: 'Genetic exceptionalism' in medicine: clarifying the differences between genetic and nongenetic tests. Ann Intern Med 2003;138:571–575.
4 Williams AG, Wackerhage H, Miah A, Harris RC, Montgomery HE: Genetic research and testing in sport and exercise science: BASES Position Stand. Leeds, British Association of Sport and Exercise Sciences, 2007.
5 Ruiz JR, Sui X, Lobelo F, Morrow JR Jr, Jackson AW, Sjostrom M, Blair SN: Association between muscular strength and mortality in men: prospective cohort study. BMJ 2008;337:a439.
6 Strittmatter WJ, Saunders AM, Schmechel D, Pericak-Vance M, Enghild J, Salvesen GS, Roses AD: Apolipoprotein E: high-avidity binding to beta-amyloid and increased frequency of type 4 allele in late-onset familial Alzheimer disease. Proc Natl Acad Sci USA 1993;90:1977–1981.

7 Human Genetics Commission: Genetic information, public consultation: Second Annual Report of the Human Genetics Commission. London, Human Genetics Commission, 2002.
8 Bray MS, Hagberg JM, Perusse L, Rankinen T, Roth SM, Wolfarth B, Bouchard C: The human gene map for performance and health-related fitness phenotypes: the 2006–2007 update. Med Sci Sports Exerc 2009;41:35–73.
9 Juvonen E, Ikkala E, Fyhrquist F, Ruutu T: Autosomal dominant erythrocytosis caused by increased sensitivity to erythropoietin. Blood 1991;78:3066–3069.
10 de la Chapelle A, Traskelin AL, Juvonen E: Truncated erythropoietin receptor causes dominantly inherited benign human erythrocytosis. Proc Natl Acad Sci USA 1993;90:4495–4499.
11 Schuelke M, Wagner KR, Stolz LE, Hubner C, Riebel T, Komen W, Braun T, Tobin JF, Lee S-J: Myostatin mutation associated with gross muscle hypertrophy in a child. N Engl J Med 2004;350:2682–2688.
12 McPherron AC, Lawler AM, Lee SJ: Regulation of skeletal muscle mass in mice by a new TGF-beta superfamily member. Nature 1997;387:83–90.
13 Mendias CL, Bakhurin KI, Faulkner JA: Tendons of myostatin-deficient mice are small, brittle, and hypocellular. Proc Natl Acad Sci USA 2008;105:388–393.
14 Myerson S, Hemingway H, Budget R, Martin J, Humphries S, Montgomery H: Human angiotensin I-converting enzyme gene and endurance performance. J Appl Physiol 1999;87:1313–1316.
15 Williams AG, Rayson MP, Jubb M, World M, Woods DR, Hayward M, Martin J, Humphries SE, Montgomery HE: The ACE gene and muscle performance. Nature 2000;403:614.
16 Zhang B, Tanaka H, Shono N, Miura S, Kiyonaga A, Shindo M, Saku K: The I allele of the angiotensin-converting enzyme gene is associated with an increased percentage of slow-twitch type I fibers in human skeletal muscle. Clin Genet 2003;63:139–144.
17 Rankinen T, Perusse L, Gagnon J, Chagnon YC, Leon AS, Skinner JS, Wilmore JH, Rao DC, Bouchard C: Angiotensin-converting enzyme ID polymorphism and fitness phenotype in the HERITAGE Family Study. J Appl Physiol 2000;88:1029–1035.
18 Yang N, MacArthur DG, Gulbin JP, Hahn AG, Beggs AH, Easteal S, North K: ACTN3 genotype is associated with human elite athletic performance. Am J Hum Genet 2003;73:627–631.
19 Niemi AK, Majamaa K: Mitochondrial DNA and ACTN3 genotypes in Finnish elite endurance and sprint athletes. Eur J Hum Genet 2005;13:965–969.

20 MacArthur DG, Seto JT, Chan S, Quinlan KG, Raftery JM, Turner N, Nicholson MD, Kee AJ, Hardeman EC, Gunning PW, Cooney GJ, Head SI, Yang N, North KN: An Actn3 knockout mouse provides mechanistic insights into the association between alpha-actinin-3 deficiency and human athletic performance. Hum Mol Genet 2008;17:1076–1086.
21 Clarkson PM, Devaney JM, Gordish-Dressman H, Thompson PD, Hubal MJ, Urso M, Price TB, Angelopoulos TJ, Gordon PM, Moyna NM, Pescatello LS, Visich PS, Zoeller RF, Seip RL, Hoffman EP: ACTN3 genotype is associated with increases in muscle strength in response to resistance training in women. J Appl Physiol 2005;99:154–163.
22 Moran CN, Yang N, Bailey ME, Tsiokanos A, Jamurtas A, Macarthur DG, North K, Pitsiladis YP, Wilson RH: Association analysis of the ACTN3 R577X polymorphism and complex quantitative body composition and performance phenotypes in adolescent Greeks. Eur J Hum Genet 2007;15:88–93.
23 Conley DL, Krahenbuhl GS: Running economy and distance running performance of highly trained athletes. Med Sci Sports Exerc 1980;12:357–360.
24 Lucia A, Olivan J, Gomez-Gallego F, Santiago C, Montil M, Foster C: Citius and longius (faster and longer) with no alpha-actinin-3 in skeletal muscles? Br J Sports Med, 2007. http://bjsm.bmj.com/cgi/content/abstract/41/9/616; retrieved 12 December 2008.
25 Mokone GG, Gajjar M, September AV, Schwellnus MP, Greenberg J, Noakes TD, Collins M: The guanine-thymine dinucleotide repeat polymorphism within the tenascin-C gene is associated with Achilles tendon injuries. Am J Sports Med 2005;33:1016–1021.
26 Williams AG, Folland JP: Similarity of polygenic profiles limits the potential for elite human physical performance. J Physiol 2008;586:113–121.
27 Roth SM: Perspective on the future use of genomics in exercise prescription. J Appl Physiol 2008;104:1243–1245.
28 Corrado D, Basso C, Pavei A, Michieli P, Schiavon M, Thiene G: Trends in sudden cardiovascular death in young competitive athletes after implementation of a preparticipation screening program. JAMA 2006;296:1593–1601.
29 Hogarth S, Javitt G, Melzer D: The current landscape for direct-to-consumer genetic testing: legal, ethical, and policy issues. Annu Rev Genomics Hum Genet 2008;9:161–182.

30 Rantanen E, Hietala M, Kristoffersson U, Nippert I, Schmidtke J, Sequeiros J, Kaariainen H: What is ideal genetic counselling? A survey of current international guidelines. Eur J Hum Genet 2008;16:445–452.

31 Osterweil N: Full court press on hoop star Curry to get DNA testing. MedPage Today, 2005. http://www.medpagetoday.com/Cardiology/Arrhythmias/1843; retrieved 7 January 2009.

32 World Anti-Doping Agency: WADA gene doping symposium reaches conclusions and recommendations. World Anti-Doping Agency, 2005. http://www.wada-ama.org/en/newsarticle.ch2?articleId = 3115229; retrieved 7 January 2009.

33 Nicolas P: Ethical and juridical issues of genetic testing: a review of the international regulation. Crit Rev Oncol Hematol 2009;69:98–107.

34 Hesketh T, Xing ZW: Abnormal sex ratios in human populations: causes and consequences. Proc Natl Acad Sci USA 2006;103:13271–13275.

Alun G. Williams, PhD
Department of Exercise and Sport Science
Manchester Metropolitan University
Hassall Road, Alsager ST7 2HL (UK)
Tel. +44 161 247 5523, Fax +44 161 247 6375, E-Mail a.g.williams@mmu.ac.uk

The Future of Genetic Research in Exercise Science and Sports Medicine

Ronald J. Trent · Bing Yu

Central Clinical School, University of Sydney, and Department of Molecular and Clinical Genetics, Royal Prince Alfred Hospital, Camperdown, NSW, Australia

Abstract

Background/Aims: Genetic research is used to identify the relative contributions made by inherent abilities (nature) versus environmental effects (nurture) in human performance. The same approach allows a better understanding of how injuries or illnesses can result from sport or physical activity. Having identified the genes involved in athletic performance, there are the intriguing possibilities of using this information for talent search, developing individualized training programs and prevention of sports-related injuries. **Methods:** There are many interacting genes involved in athletic performance. This class of genes is often described as 'complex' and the mode of inheritance is called 'multifactorial'. Discovery of these genes is difficult using the conventional case control (association) studies. Recent genomic-based developments allowing high throughput SNP analysis are very promising. Potentially more exciting is the availability in the near future of cheaper and faster whole-genome sequencing technologies. **Results:** Genetic research in exercise science has produced a lot of data including the ability to draw a human exercise gene map. However, progress at the genetic level has been slow because gene-based association studies are not powerful enough to detect multiple small but cumulative gene effects. In future, the more efficient genomic-based research approaches will accelerate the finding of 'sports genes'. Data generated will be enormous, making it essential to have a direct link between the laboratory researcher and bioinformatics expertise. **Conclusion:** Genetics research has moved to the genomics era, i.e. the simultaneous testing of multiple genes is now possible.

Copyright © 2009 S. Karger AG, Basel

From Genetics to Genomics

The increasing demands required for success in sport at the elite level are usually approached through better talent search, more effective training regimens and improved infrastructural support. However, the developments achievable are usually incremental with the potential for quantum leaps only likely to come from cutting-edge research.

One major research interest is in the area of talent search. There are two main goals: (a) How can the most talented athletes be detected at an early age? (b) What is the training potential for the individual, i.e. has the person with identified athletic ability reached his or her peak or is further improvement possible following training? Genetics-based research promises a lot in both of these quests.

Insight into how our genes work in the disease process provides a pathway to more effective treatments as well as earlier prevention strategies. Indeed, it is well accepted in clinical practice that knowledge of mutations in genes can be used to predict many years in advance, even before symptoms are apparent, that an individual will develop a disease. This is possible because a unique feature of a DNA (genetic) test comes from its *predictive* potential.

While inherently attractive, the applications of genetic knowledge in exercise science as well as the prevention of injuries in sports medicine remain elusive. Why? The genetics-in-disease paradigms are presently restricted to mendelian type genetic defects. These usually involve mutations in single genes having a major effect on protein output or structure and the result is a genetic disorder. Most mendelian type genetic disorders are relatively rare and their effects can be tracked through family pedigrees. In contrast, the more important public health problems in many communities, e.g. ageing, dementia, cardiovascular disease and most forms of cancer, are inherited in more subtle ways with the likelihood that there are cumulative effects from many genes contributing to the one disorder, and the effects of these genes can be modulated by interactions with the environment. In these so-called 'multifactorial genetic disorders', it is not possible to draw a pedigree and so trace the small but additive effects of multiple genes.

Normal traits, and included here would be the genes involved in human performance, fall into the class of multifactorial inheritance. The traditional genetic-based strategies to discover genes for multifactorial inheritance involve case-control studies (called association studies in genetics) that have many inherent difficulties including the high probability that false-positive or false-negative results will occur. There are a variety of reasons for failures in association studies including: (a) accuracy of the phenotype, (b) small numbers of subjects tested, (c) small numbers of gene markers. These will invariably affect the power in terms of study design [1]. (d) The candidate gene approach is itself a source of error with many unsuspected but important genes likely to be missed.

In contrast, the genomic-based whole-genome association studies allow multiple DNA markers (usually SNPs – single nucleotide polymorphisms) to be studied and these span the whole genome. Provided sufficient numbers of subjects are enrolled, it is highly likely that the whole genome association approach, because of the extensive coverage coming from the large number of SNPs used, will provide some important 'hits'. These will or will not confirm the earlier genetic association studies in sports research and will identify additional relevant loci and genes.

During 2006–2007, there was a significant increase in the number of genes for complex genetic traits discovered using the more sophisticated whole-genome association strategy. This was possible because of new developments in typing for DNA polymorphisms known as SNPs. Today, one can select a large number of SNPs based on the HapMap knowledge of linkage disequilibrium and type these DNA markers more

Table 1. Comparison of the traditional genetic association study and the genome-wide association

Feature	Existing genetic association	Genome-wide association
Genetic markers (SNPs)	Small number of SNPs – usually less than 20 and selected individually	Large number of SNPs, e.g. 500,000 – commercially available
Cases and controls	100–300 (individual research groups)	1,000–3,000 (these numbers possible with international consortia)
Population ethnicity and stratification	Self-reported and so lacking objective control (leading to the risk of population admixture)	Well controlled with a subset of analyzed SNPs (avoiding the population admixture issue)
Genes to be tested	One or a few candidate genes	All genes are tested without any assumed hypothesis
Effect sizes (odds ratio)	>5.0 (less likely to be true in complex trait)	1.5–2.0 (more likely to be identified in complex traits)

effectively across the whole of the human genome [1]. This approach has required the formation of international consortia to ensure that the numbers of subjects tested (in the thousands) are a magnitude higher than the more basic gene association studies (in the tens to hundreds of subjects; table 1).

Exercise Science

Talent Search

The many genes already implicated in human athletic ability have been captured in the form of a human gene map for performance and health-related fitness phenotypes [2]. The rapid developments in this area are illustrated by the 2006–2007 update which lists 214 autosomal gene sites and quantitative trait loci, an additional seven on the X chromosome and 18 mitochondrial genes. In contrast, the 2000 version of the map had only 29 identified loci. This catalogue is based on a review of the literature. Therefore, it will include many studies that remain to be confirmed.

The next level of sophistication for the above map would be verification of the various loci with a meta-analysis approach or more sophisticated strategies. The latter will come from whole genome association studies. Although still expensive (because of the large number of SNPs and samples used), the power generated is impressive, and so able to detect even small cumulative gene effects, i.e. quantitative trait loci. The attraction of the whole genome association approach will ensure that researchers pool their most important resource, i.e. DNA from elite athletes, by forming international consortia (table 1).

Individualized Training Programs
What is the best training program for a particular sport, or in the modern genetics jargon, can training programs be personalized based on the individual's intrinsic (i.e. genetic) strengths and weaknesses? Related to this is the dilemma facing national Sports Institutes when burnout occurs as a consequence of the training program. As well as being a personal loss to an individual who has dedicated a significant portion of his or her life to sport, burnout represents a significant financial loss to the sporting institutes.

This aspect of exercise science will be achievable in the longer term because the athletes can act as their own controls in comparative studies. However, it will continue to progress slowly until more is known about the genes and their interactions with other genes and the environment.

Sports Medicine

Preventing Sports-Related Medical Complications
Athletes will invariably suffer injuries as part of their training activities or during a sporting event. Most of these are assumed to be caused by the sport undertaken, but is this the complete explanation? The question now being asked is whether some individuals are more prone to injury because of their genetic makeup.

An example would be injuries to the Achilles tendon. Table 2 shows a number of factors that have been associated with Achilles tendon pathology [3]. A study in 2004 suggested that an individual with an Achilles tendon rupture had a significantly higher risk of getting a contralateral rupture as well [4]. To a geneticist, a bilateral problem suggests an underlying genetic factor is involved. Hence, not surprisingly, genetic studies have been undertaken to determine if Achilles tendon problems may also have an underlying genetic abnormality. The candidate gene chosen was *COL5A1* gene which is important for collagen formation. Mutations in this gene lead to a severe connective tissue disorder known as the Ehlers-Danlos syndrome. But would more subtle changes in this genes expression impair collagen and so predispose to problems with the Achilles tendon which is predominantly composed of collagen? The results showed changes at the 3′ end of the gene that appeared to correlate with the development of Achilles tendon pathology [3].

Like all genetic association studies, this is only a preliminary finding and would need verification, particularly in relation to functional studies. However, it illustrates the potential of this type of genetic research because confirmation of this genetic marker would enable the identification of those at greater risk for this sports-related injury, and so more effective prevention measures could be put in place to avoid it.

Sudden Cardiac Death during Exercise and Sport
A very tragic event is the sudden and unexpected death of a child or an otherwise healthy young adult during a sporting event, or while undertaking what appears to be routine

Table 2. Causes and contributing factors to Achilles tendon damage [3]

Intrinsic factors	Age, gender, body weight, vascular perfusion, nutrition, anatomical variants, joint laxity, muscular weakness or imbalance, the use of certain drugs, systemic disease and a history of previous injury.
Extrinsic factors	Type of activity, occupation, training, physical load, shoes, environmental conditions.

exercise such as jogging. Apart from the emotive issue, the death raises two important questions: (a) Could this have been prevented if the individual had been known to be at risk? (b) Is this a genetic problem, and if so are other members of the family at risk?

A number of important genetic disorders can first manifest during competitive sports, and unfortunately these presentations can be a sudden cardiac death. Examples include Marfan syndrome (e.g. dissection of the aorta) and familial hypertrophic cardiomyopathy (e.g. arrhythmia). Because of the predictive nature of DNA testing, it should be possible, in theory, to undertake the appropriate tests for the above two genetic disorders to determine if an individual is at risk. In practice though, the underlying mutations for these disorders are either family specific and/or involve multiple or large genes, and so the current genetic DNA testing approaches are not sophisticated enough to allow a definitive screening program to be instituted. However, this will change as genomic technologies evolve and allow multiple genes to be tested and/or DNA sequencing becomes cheaper and faster, allowing individual genes, no matter how large they are, to be sequenced (this approach is discussed at the end of this paper). In other words, the developments as we move from genetics to genomics will ensure that no matter how large the gene or how family specific is the mutation, it will be possible to screen individuals in a cost-effective way to identify potential mutations in all genes known to cause cardiovascular problems such as the ones described earlier.

With this technology and information available to both the individual and the sporting institute, will it be possible to prevent these tragic deaths? To answer this question will require more (some would say urgent) research. The challenges involved are many and are summarized in the editorial by Kjaer [5].

Ethical Social Legal Issues

There are a number of opportunities opened through predictive DNA testing in the sporting area. Nevertheless, the technology is still being developed that will allow better ways in which to use DNA tests. The ultimate of these would be a whole genome sequence – discussed towards the end of this chapter. In the meantime, some questions are worth considering. What duty of care is required of a sporting organization in terms of protecting its athletes? Presently, it would be reasonable to include as part of a fitness workup of new recruits that certain physical characteristics might be indicators of health risk during competitive sports, e.g. an individual who is very tall and/or has features of Marfan syndrome. Individuals with these features should be referred for a formal clinical assessment. In this circumstance, the finding of an unexplained cardiac murmur or an aortic abnormality on imaging, confirms a high probability of Marfan syndrome and therefore significant risks associated with competitive sports.

On the other hand, whether a physical finding such as 'tallness' should automatically lead to a DNA test to screen for Marfan syndrome is being debated in terms of its feasibility and its cost versus benefit. Perhaps more problematic is what will the sporting organization do if it found one of its members had a mutation in the *FBN1* gene (the gene for Marfan syndrome) but no other clinical features? The DNA *predictive* test for Marfan syndrome is not a straightforward test because of the issue of variable severity, i.e. an *FBN1* gene mutation alone will not necessarily identify the severity of the disorder. What duty of care does the sporting body have to the athlete in this circumstance? Would it fail its duty of care if it allowed the individual to continue to train at the institute? On the other hand, would excluding the athlete because of the perceived health risk comprise unlawful discrimination?

Gene Doping

Invariably when the subject of genetics and sports comes up, the topic of conversation eventually shifts to gene doping. This is an unfortunate connection because the real challenges of genetics in sport, apart from the difficult research questions, focus around the use of genetic information and how this needs to be protected and used appropriately. Gene doping involves the manipulation of DNA, in particular gene transfer (also called gene therapy), to deliver a banned substance or to enhance or modify a natural gene product. Gene doping is no different to drug cheating or blood doping, and the fact that it involves DNA technology is incidental. In medical practice, gene transfer is still an experimental procedure with few successful results to date. In addition, the approach is risky and one of the complications already seen is interference with the function of a normal gene resulting in leukemia. Hence, gene doping is particularly inappropriate because it is banned by the World Anti-Doping Agency, and it involves an experimental procedure which in itself poses significant health risks to the individual [6]. What should not be forgotten in the emotive discussion around gene transfer and human performance is that in the longer term if gene transfer is able to be used effectively in the clinical setting, it might have an important application in treating some sport-related injuries [6].

Challenges for the Future

Frameworks for Research
All research conducted in human performance will go through the usual regulatory requirements, particularly assessment by a human research ethics committee,

to ensure that the appropriate safeguards are in place. Privacy issues are particularly sensitive when it comes to high-profile elite sportsmen and sportswomen because of the interest shown by the media in these individuals. In 2003, the United Nations Educational, Scientific and Cultural Organization adopted the International Declaration on Human Genetic Data [7]. This recognized that human genetic data have a *special status* because they:

- can be predictive of genetic predispositions involving individuals
- may have a significant impact on the family, including offspring, extending over generations and in some instances on the whole group to which the person concerned belongs
- may contain information the significance of which is not necessarily known at the time of the collection of biological samples, and
- may have cultural significance for persons or groups.

Not surprisingly, the potential misuse of genetic information in sport has attracted the attention of government, sports administrators and ethicists. One example is the extensive joint enquiry by the Australian Law Reform Commission and the National Health and Medical Research Council's Australian Health Ethics Committee into broad aspects of human genetics in Australia covering medical to non-medical uses of genetic information. The 2003 final report titled *'Essentially Yours: the protection of human genetic information in Australia'* [8] made two recommendations specifically directed to genetics and sport (see box).

> **Recommendation 38-1:** The Australian Sports Commission should monitor the use of genetic testing and genetic information for identifying or selecting athletes with a view to developing policies and guidelines for sports organizations and athletes.
> **Recommendation 38-2:** the Australian Sports Commission should develop policies and guidelines for sports organizations and athletes on the use of genetic information in relation to predisposition to sports-related illness or injury.

Direct-to-Consumer DNA Testing

A growth industry in the past few years has been direct-to-consumer (DTC) DNA testing. This describes the availability of DNA tests direct to the public without the involvement of health professionals. DTC DNA testing is offered for a number of 'products': (a) DNA genetic tests for medical conditions, (b) quasi-medical DNA tests such as dermatogenetics, i.e. based on an individual's own DNA information companies claim they can recommend appropriate creams or therapies that will ensure a youthful skin appearance. (c) Lifestyle DTC DNA tests – DNA markers are used for various nonmedical purposes such as genealogy searches or, in the context of sporting performance, the potential for an *individual* to excel in sport.

An example of a sports-related DNA test is the *ACTN3* gene test which is available through Genetic Technologies in Australia, and recently it has been licensed to a US-based DNA testing laboratory [9]. The scientific basis for linking *ATN3* to athletic performance was first established in 2003 and still remains the subject of interest and research [10]. What remains to be determined is the benefit (if any) this DNA test has for predicting an *individual's* athletic ability, a question which is particularly relevant because the test is most likely to be used on a child.

Unfortunately, the rapid expansion in the number of DTC DNA testing laboratories, particularly those that are difficult to regulate because they are located offshore, and the broad claims made by these companies have the potential to hold back the more legitimate uses of DNA (genetic) testing. The negative effects on DNA testing by inappropriate use of DTC DNA testing, particularly when it comes to predictive and personalized genetics, is illustrated by a recent report from the US Government Accountability Office highlighting how four web-based nutrigenetics companies misled consumers [11].

New Technologies

The 'greatest transformational aspect of the Human Genome Project has been not the sequencing of the genome itself, but the resultant development of new technologies' [12]. The first human genome sequenced in 2003 is estimated to have cost about USD 3 billion. The next human genome sequenced in 2007 (that of Prof. James Watson) is said to have cost USD 2 million. Today, a human genome sequence can cost about USD 350,000 [13]. The next goal for a complete human gene sequence is the USD 1,000 price tag. Once this is achievable, the options for research studies will greatly expand with the sequencing of individuals providing a glut of comparative DNA data that will require sophisticated bioinformatics-based analysis.

New Paradigms

An important scientific fact emerging from the Human Genome Project is we have a simplistic understanding of what is a gene. This is well illustrated by the 2007 mapping of the pinot noir grape genome [14]. It showed this grape had slightly more genes than humans i.e. 30,000 versus 25,000. So what is it that makes humans more sophisticated than the pinot noir grape? 'Genes' per se cannot be the entire explanation, and so not surprisingly in terms of human performance, researchers will need to look beyond what is the traditional gene including:

Copy Number Variants. Although SNPs are the focus of current genetic research, there are other forms of variation within the genome, particularly the copy number variants. Gross structural changes leading to an increase or decrease in chromosome or chromosomal regions have been known for some time to produce genetic disease. However, much smaller copy number changes in the genome are also present with one estimate being that there are more than 1,400 copy number variable regions representing over 12% of the genome [14]. This remains an unexplored option in terms of how genes might influence human performance.

Epigenomics. Traditionally, the focus in genetic research is the gene and its DNA sequence. However, it is clear that gene expression can be altered without changing the actual sequence by chemically modifying the base cytosine or altering the proteins associated with DNA and chromosomes. These are known as epigenetic changes.

Today, there is evidence that sporadic tumors demonstrate epigenetic effects, i.e. either genes such as tumor suppressor genes are inappropriately turned off through epigenetic changes, or genes that should not be functioning like oncogenes, are inappropriately turned on because of loss of epigenetic control [14].

RNA is now seen as an important player in gene function. In particular, the potential for a new species of small (micro-) RNAs (miRNAs) to regulate DNA provides an additional epigenetic mechanism for gene regulation as well as protein diversity [14]. Do miRNAs explain why our genome in terms of gene number is similar to the pinot noir grape but the human proteome is considerably more complex? Time will tell, but the 2006 Nobel Prize in Physiology or Medicine for a species of miRNAs would suggest that research into sport will need to consider seriously the role played by epigenetics, and in particular miRNAs, in gene function.

References

1 Altshuler D, Daly MJ, Lander ES: Genetic mapping in human disease. Science 2008;322:881–888.
2 Bray MS, Hagberg JM, Perusse L, Rankinen T, Roth SM, Wolfarth B, Bouchard C: The human gene map for performance and health-related fitness phenotypes: the 2006–2007 update. Med Sci Sports Exerc 2009;41:34–72.
3 Mokone GG, Schwellnus MP, Noakes TD, Collins M: The COL5A1 gene and Achilles tendon pathology. Scand J Med Sci Sports 2006;16:19–26.
4 Aroen A, Helgo D, Granlund OG, Bahr R: Contralateral tendon rupture risk is increased in individuals with a previous Achilles tendon rupture. Scand J Med Sci Sports 2004;14:30–33.
5 Kjaer M: Sudden cardiac death associated with sports in young individuals: is screening the way to avoid it? Scand J Med Sci Sports 2006;16:1–3.
6 Trent RJ, Alexander IE: Gene therapy in sport. Br J Sports Med 2006;40:4–5.
7 UNESCO's international declaration on human genetic data. http://unesdoc.unesco.org/images/0013/001342/134217e.pdf; accessed 19.12.2008.
8 The Australian Law Reform Commission website providing the *Essentially Yours* report as well as the Commonwealth Government's response to this report. http://www.alrc.gov.au/inquiries/title/alrc96/index.htm; accessed 04.06.2009.
9 Genetic Technologies Ltd: Providing direct to consumer DNA testing for athletic ability through the ACTN3 DNA test. http://www.gtg.com.au; accessed 19.12.2008.
10 Yang N, MacArthur DG, Gulbin JP, Hahn AG, Beggs AH, Easteal S, North K: ACTN3 genotype is associated with human elite athletic performance. Am J Hum Genet 2003;73:627–631.
11 Report from the US Government Accountability Office dealing with DTC DNA tests purchased from four web sites. http://www.gao.gov/products/GAO-06-977T; accessed 19.12.2008.
12 Kahvejian A, Quackenbush J, Thompson JF: What would you do if you could sequence everything? Nature Biotech 2008;26:1125–1133.
13 Knome Inc: First personal genomics company to offer complete genome sequencing and analysis services for private individuals. http://www.knome.com/about/faqs.html#cost; accessed 19.12.2008.
14 The French-Italian Public Consortium for Grapevine Genome Characterization: The grapevine genome sequence suggests ancestral hexaploidization in major angiosperm phyla. Nature 2007;449:463–417.

Prof. Ronald J. Trent
Department of Molecular and Clinical Genetics
Royal Prince Alfred Hospital, Missenden Rd.
Camperdown, NSW 2050 (Australia)
Tel. +61 2 9515 7514, Fax +61 2 9550 5412, E-Mail rtrent@med.usyd.edu.au

Author Index

Ahmetov, I.I. 43

Beunen, G.P. 28
Brutsaert, T.D. 11

Collins, M. VII, 136

Franks, P.W. 110

Garton, F. 88
Gibson, W.T. 1
Gross, G. 150

Hoffmann, A. 150

Malina, R.M. 28

North, K. 88

Onywera, V.O. 102

Parra, E.J. 11
Peeters, M.W. 28
Pomeroy, J. 110

Raleigh, S.M. 136
Rogozkin, V.A. 43

Söderberg, A.M. 110

Thomis, M.A.I. 28
Trent, R.J. 187

Wackerhage, H. 176
Wells, D.J. 166
Williams, A.G. 176
Woods, D. 72

Yang, N. 88
Yu, B 187

Subject Index

ABCC8, gene-lifestyle interactions 121
ACE, *see* Angiotensin-converting enzyme
Actinin, *see* ACTN3
ACTN3
   alleles
      African runners 104
      endurance athletes 95, 96
      genetic testing 180, 181
      power athletes 53, 54, 63
   deficiency
      effects on muscle
         fiber contraction 97
         mass and fiber type 98, 99
         metabolism 97
         strength and power 97, 98
      frequency 90, 91
      null allele and positive natural selection 91
      public health implications 99, 100
   genotype in elite athletes 92–95
   knockout mouse studies 95
   skeletal muscle role 88–90
ADAMTS, polymorphisms in soft tissue injuries 143, 144
Adenosine demaninase, *see* AMPD1
ADRA2A, allele in endurance athletes 46
ADRB2
   allele in endurance athletes 46, 47
   gene-lifestyle interactions 121, 123, 127
ADRB3, gene-lifestyle interactions 119, 123, 127
Adrenergic receptors, *see* ADRA2A; ADRB2
Aerobic capacity
   heritability 12, 13, 35–37
   low birthweight effects 20
Aerobic training, genetics and response 38, 56, 57, 59–62
African runners
   diet and lifestyle 105, 106
   genetics 102–104
   motivation 105, 108
   parental influences 107
   talent identification and nurturing 107
   traditional facilities and advantages 106
Allele, definition 2
AMPD1, allele in endurance athletes 47, 59
Amylase, copy number variants 14
Anaerobic training, *see* Strength training
Angiotensin-converting enzyme (ACE), gene polymorphisms
   African runners 103
   cross-sectional studies 73–77
   endurance athletes 44, 46, 55, 56, 73–77
   endurance training response 57, 59
   gene-lifestyle interactions 127
   genetic testing 180
   high-altitude performance studies 79
   left ventricular hypertrophy 43, 44
   mechanisms of performance effects
      high altitude 82
      sea level 79–82
   overview 13, 15
   power athletes 53, 62, 63, 78, 79
   prospective studies 78, 79
   prospects for study 83
   types 72, 73
Anterior cruciate ligament, *see* Ligament
ApoE, *see* Apolipoprotein E
Apolipoprotein E (ApoE)
   alleles and training effects 60
   gene-lifestyle interactions 127
Association study, definition 5
Athletic ability
   environmental factors 16–21
   genetic factors 12–16

ATP1A2, alleles and training effects 60

BDKRB2, allele in endurance athletes 47, 48, 55
Birthweight
    aerobic capacity effects 20
    bone mineral density effects 19
    lean body mass effects 18
    muscle strength effects 20
    obesity effects 18–21
    sleeping metabolic rate effects 21
Blood pressure recovery, genetics 57
BMI, see Body mass index
BMPs, see Bone morphogenetic proteins
Body mass index (BMI), heritability 31
Bone mineral density (BMD), low birthweight effects 19
Bone morphogenetic proteins (BMPs), regenerative medicine for soft tissue injury 160, 161
Bradykinin receptor, see BDKRB2

Calcineurin, see PPP3R1
Candidate gene
    association study 5, 6, 13, 188
    definition 5
Cartilage
    injury 152
    regenerative medicine 159, 160
CKM, see Creatine kinase
CNV, see Copy number variant
COL1A1, polymorphisms in soft tissue injuries 139–141
COL5A1, polymorphisms in soft tissue injuries 141, 142
COL12A1, polymorphisms in soft tissue injuries 142, 143
COL14A1, polymorphisms in soft tissue injuries 142, 143
Complex trait, characteristics 3
COMT, gene-lifestyle interactions 127
Copy number variant (CNV)
    assessment 8, 15
    overview and significance 7, 8, 14
Creatine kinase (CKM), alleles and endurance training effects 60
CYP19A1, gene-lifestyle interactions 127

DNA, structure and variants 2

Endurance training, candidate genes for response 56, 57, 59–62

EPAS1, allele in endurance athletes 48
Epigenomics, prospects for study 195
EPO, see Erythropoietin
Erythropoietin (EPO)
    gene doping 167, 168
    receptor mutation and athletic performance 179

Fetal programming
    athletic ability 16–18
    conceptual framework 16, 17
FGF, see Fibroblast growth factor
Fibroblast growth factor (FGF), regenerative medicine for soft tissue injury 158
FTO, gene-lifestyle interactions in obesity 113, 117, 123, 125

GABPB1, alleles and endurance training effects 60
Gene doping
    candidate genes for athletic performance 167, 168
    concept 166, 167
    ethics 192, 193
    injury recovery improvement 168, 169
    practical considerations 169–171
    prospects 172, 173
Gene-lifestyle interaction
    clinical applications 115
    definition 112
    history of study 112, 113
    mechanisms 113
    obesity and table of observational studies 115–131
    overview 110, 111
    prospects for study 131, 132
Gene mapping, overview 3, 4
Gene therapy
    delivery 169–171
    regenerative medicine 155, 156
Genetic testing
    commercial systems 176, 177
    differences between genetic and non-genetic performance tests 177–179
    direct-to-consumer testing 193, 194
    ethics 183, 184
    predictive potential 188
    requesting individuals 182–184
    targets for performance testing 179–182
Genome
    coding regions 8

sequencing advances and prospects for individuals 194
Genome-wide association study (GWAS)
  advantages and limitations 6, 7, 14, 15
  comparison with other studies 188, 189
  definition 5
Genotype, definition 3
G-protein-coupled receptor 10 gene-lifestyle interactions in blood pressure 115, 116
GWAS, see Genome-wide association study

Heart rate recovery, genetics 57
Hemoglobin, gene alleles and endurance training effects 61
Heritability
  assessment 29, 30
  definition 12
  prospects for study 39, 40
HIF1, alleles and endurance training effects 61
HIF1A, allele in power athletes 54
Hypoxia-inducible factor, see HIF1A

IGF1, see Insulin-like growth factor-1
Imprinting, gametes 16
Individualized training, genetics impact 190
Insulin-like growth factor-1 (IGF-1)
  alleles and power training effects 63
  gene doping 168
  regenerative medicine for soft tissue injury 158, 160
Interleukin-15, alleles and power training effects 63, 64

KCNJ11, E23K variant and metformin treatment in diabetes risk reduction 115
Kenya, see African runners

LBM, see Lean body mass
Lean body mass (LBM), low birthweight effects 18
LEPR, gene-lifestyle interactions 125, 129
Ligament
  injury
    ADAMTS polymorphisms 143, 144
    chromosome 9 gene polymorphisms 141, 142
    classification 137
    COL1A1 polymorphisms 139–141
    frequency 137
    matrix gene polymorphisms 142, 143
    matrix metalloproteinase polymorphisms 143, 144
    mechanism 38, 139
    prospects for study 144, 145
    treatment 153, 154
  regenerative medicine, see Soft tissue injury
  structure 137, 138
Locus, definition 4
LPL, gene-lifestyle interactions 125

Marfan syndrome, genetic testing 192
Matrix metalloproteinases (MMPs), polymorphisms in soft tissue injuries 143, 144
Mendelian trait, characteristics 3
Mesenchymal stem cell (MSC), regenerative medicine for soft tissue injury 156, 157, 159, 161
Mitochondrial DNA haplogroups
  African runners 104
  endurance athletes 48
Mitochondrial transcription factor A, see TFAM
MMPs, see Matrix metalloproteinases
Motor performance, heritability 34, 35
MSC, see Mesenchymal stem cell
Muscle strength
  heritability 32–34
  low birthweight effects 20
Mutation, definition 2–3
Myostatin, mutation and athletic performance 179, 180

Nerve growth factor (NGF), regenerative medicine for soft tissue injury 158
NFATC4, allele in endurance athletes 48, 49
NGF, see Nerve growth factor
Nigeria, see African runners
Nitric oxide synthase, see NOS3
NOS3, allele in endurance athletes 49
NRF1, alleles and endurance training effects 61
NRF-2, alleles and endurance training effects 60

Obesity
  gene-lifestyle interaction 115–131
  low birthweight effects 18–21

Peak height velocity (PHV), heritability 32
Peroxisome proliferators-activated receptor, see PPARA; PPARD; PPARG; PPARGC1A; PPARGC1B
Phenotype, definition 3

PHV, *see* Peak height velocity
Physique, heritability 31
PLA2G7, gene-lifestyle interactions 123, 129
Power training, *see* Strength training
PPARA
   allele in endurance athletes 49, 50
   allele in power athletes 54, 55
   gene-lifestyle interactions 125
PPARD
   allele in endurance athletes 50, 61, 62
   gene doping 168
   gene-lifestyle interactions 127
PPARG
   allele in power athletes 55
   gene-lifestyle interactions 119
PPARGC1A
   allele in endurance athletes 50, 62
   gene-lifestyle interactions 121
PPARGC1B, allele in endurance athletes 51
PPP3R1, allele in endurance athletes 51

Qualitative trait, characteristics 3
Quantitative trait, characteristics 3

Single nucleotide polymorphism (SNP)
   definition 2
   frequency 8
   genotyping 4, 5, 188
   nomenclature 4
Skill acquisition, genetics 38, 39
SLC2A2, gene-lifestyle interactions 121
Sleeping metabolic rate (SMR), low birthweight effects 21
SMR, *see* Sleeping metabolic rate
SNP, *see* Single nucleotide polymorphism
Soft tissue injury, *see also* Ligament; Tendon
   cartilage injury 152
   lesion types 151, 152
   prevention 190
   regenerative medicine
      animal studies 158–162
      cartilage 159, 160, 163
      exercise benefits 157, 158
      gene therapy 155, 156
      growth factor delivery 154, 155
      human studies 162, 163
      muscle 158, 159, 163
      overview 151
      stem cell therapy 156, 157
      tendon 160–163
Somatotype, heritability 31

Stature, heritability 30, 31
Stem cell therapy, regenerative medicine 156, 157
Strength training, genetics and response 37, 62–64
Sudden cardiac death, genetic screening 190

Tenascin-C, *see* TNC
Tendon
   injury
      ADAMTS polymorphisms 143, 144
      causes in Achilles tendon 190, 191
      chromosome 9 gene polymorphisms 141, 142
      classification 137
      COL1A1 polymorphisms 139–141
      frequency 137
      matrix gene polymorphisms 142, 143
      matrix metalloproteinase polymorphisms 143, 144
      mechanism 138, 139
      prospects for study 144, 145
      treatment 153, 154
   regenerative medicine, *see* Soft tissue injury
   structure 137, 138
TFAM, allele in endurance athletes 51
TNC
   genetic testing 181
   polymorphisms in soft tissue injuries 141, 142
Tumor necrosis factor-α, alleles and power training effects 64
Twin studies, overview 29, 30

UCP1, gene-lifestyle interactions 127
UCP2
   allele in endurance athletes 52
   gene-lifestyle interactions 119
UCP3
   allele in endurance athletes 52
   gene-lifestyle interactions 119, 121
Uncoupling proteins, *see* UCP2; UCP3

Vascular endothelial growth factor, *see* VEGFA
VEGFA, allele in endurance athletes 52, 53, 62

Y-chromosome, allele in endurance athletes 53